Ice Age 2050s Certified
Illustrated Science Exploration by Rolf A. F. Witzsche

© Text Copyright Rolf A. F. Witzsche 2019
all rights reserved

This book contains the transcript with images of the exploration video with the above title:
see: http://www.ice-age-ahead-iaa.ca/

Lead in:

We are looking at 3 distinct processes of the Ice Age unfolding in the escalating climate change on the road to the Ice Age.

Phase 1: Cosmic-ray induced global cooling and climate collapse.
Phase 2: The dimming of the Sun starts when the solar wind stops.
Phase 3: Primer Fields collapse - hibernating Sun starts the Ice Age.

The Ice Age Challenge is not of a type that can be responded to reactively, similar to nuclear war. If we delay our response till it happens, it will be too late. Nor is the actual Phase 3 start-up of the Ice Age of critical importance. The critical phase is Phase 1, the phase of the climate collapse that devastates agriculture as the Earth gets colder and drier, year after year, for the next 15 years. If we rise to the challenge and build us a climate independent New Word with technological infrastructures, such as indoors agriculture afloat in the tropics, we will create ourselves a secure future, the brightest ever, versus having none. But will we do it?

Table of Contents

The climate of our world is continuously changing 8

We have seen warm periods and cold periods 9

Climate is a critical factor for us all 10

The Little Ice Age a period of mass starvation 11

We are now in a period of global cooling 12

Crop yields are diminishing 13

World grain production has already dropped 14

The entire world is affected by the changing global climate 15

Food and climate are presently linked 16

The large-scale infrastructures can be build that secure our future 17

Barriers set up against science, to keep it small 18

Breaking through long-standing barriers 19

H. G. Wells wrote a novel, The Time Machine 20

H. G. Wells was saying to the elite 21

It was in the 1920s, when the orbital cycles theory was invented 22

Milankovitch, a mathematician, combined the orbital variations 23

The total solar irradiation of the Earth never varies 24

Total energy received from the Sun remains the same 25

Johannes Kepler had made those basic recognitions 26

Nor did Milankovitch say a word about the sunspot cycles 27

Proof that the Sun is the Earth's climate master 28

The Carbon-14 measurements 'certify' 29

But do the Carbon-14 measurements, by themselves, prove that the cosmic-ray flux did come from the Sun? . 30

The needed proof was delivered by Berillium-10 31

The Berillium-10 measurements yield higher-resolution results 32

The berillium-10 ratios follow the solar cycles closely 33

But do these measurements proof that changing solar activity is causing the ice ages? No 34

To look deep into interglacial climate history 35

The weakening of the warm climate that has been measured 36

Certainty that cosmic factors affect our climate 37

Ice Ages do not happen gradually 38

Ice Ages result from dramatic oscillations of the climate 39

Glaciation , like a person falling off a cliff, ... 40

Cooling had been up to 40 times deeper than during the Little Ice Age 41

The comparison gives us a sense of the scale of the climate change 42

That 80% less precipitation occurs during glacial times 43

Ice ages are essentially digital in nature 44

Hope and pray for a soft landing 45

What type of cushion for a soft landing to happen? 46

The steep climb out of the glacial climate 47

Contained within the narrow band of the interglacial climate 48

When we fall off the cliff to glacial times 49

Soft landing for dropping off the high cliff of interglacial 50

A more truthful recognition of the nature of the Sun is required 51

The Plasma Cause 52

Evidence for the plasma cause 53

Kepler was amazed by the geometric progression 54

On the mechanistic platform, no cause can be found 55

Kepler was puzzled by the geometric harmony 56

Kepler saw the geometric progression of node rings 57

The link with plasma evidence 58

Plasma in Space 59

The Sun itself is a body of evidence 60

Plasma is made up of electric particles. In their motion, magnetic principles apply 61

In space plasma flows in streams 62

When the magnetic fields tangle up 63

David LaPoint named the magnetic fields, The Primer Fields 64

He replicated in laboratory experiments 65

Continued weakening will collapse the the Primer Fields 66

I have explored the dynamics extensively 67

Another mayor contributor to the Primer Fields theory 68

Another major contributor was the NASA and ESO Ulysses satellite 69

The voids in the solar wind over the poles of the Sun 70

The Ulysses mission certified the Plasma Sun 71

Ulysses about the solar wind 72

The solar wind no longer occurs 73

While the solar wind flows 74

The cut-off of the solar wind is critical as an indicator 75

Once the Primer Fields have vanished we are only half-way through Phase 1 of the climate collapse 76

Ever-larger temperature extremes as the greenhouse is diminishing 77

The entire climate-change process by solar cosmic-ray flux 78

Cosmic-ray flux is continuously increasing 79

The critical timing 80

Throughout Phase I in the boundary zone 81

Solar light and heat radiation nearly constant 82

Only when we get into Phase 2 83

The available growing season is barely sufficient 84

Large areas will become unsuitable for food production 85

Could shrink the food growing area in Canada's grain belt 86

The Ice Age Challenge like Nuclear War 87

What happens after Phase 1 is of no great significance 88

The Universe won't let us wait.. 89

The Ice Age start-up will happen 90

The global warming pulses have been getting smaller 91

The short-term oscillations too, have diminished 92

The Universe won't let us wait.. 93

The Ice Age start-up will happen 94

The global warming pulses have been getting smaller 95

The short-term oscillations too, have diminished 96

Diminished in a near-geometric progression 97

The solar heartbeat is slowing 98

The electric resonance gives the Earth its brief interglacial holidays 99

If the resonance effects didn't happen none of us would exist.. 100

Humanity doesn't exist to merely live 101

The New World infrastructures cannot be build on a platform of failures 102

The only foundation on which we can succeed, is the power of our humanity 103

Our historic achievements should be seen as a foundation to build on 104

If this is the direction in which we seek our future... 105

More Illustrated Science Books by Rolf A. F. Witzsche 106

The climate of our world is continuously changing

It is increasingly evident that the climate of our world is continuously changing.

We have seen warm periods and cold periods

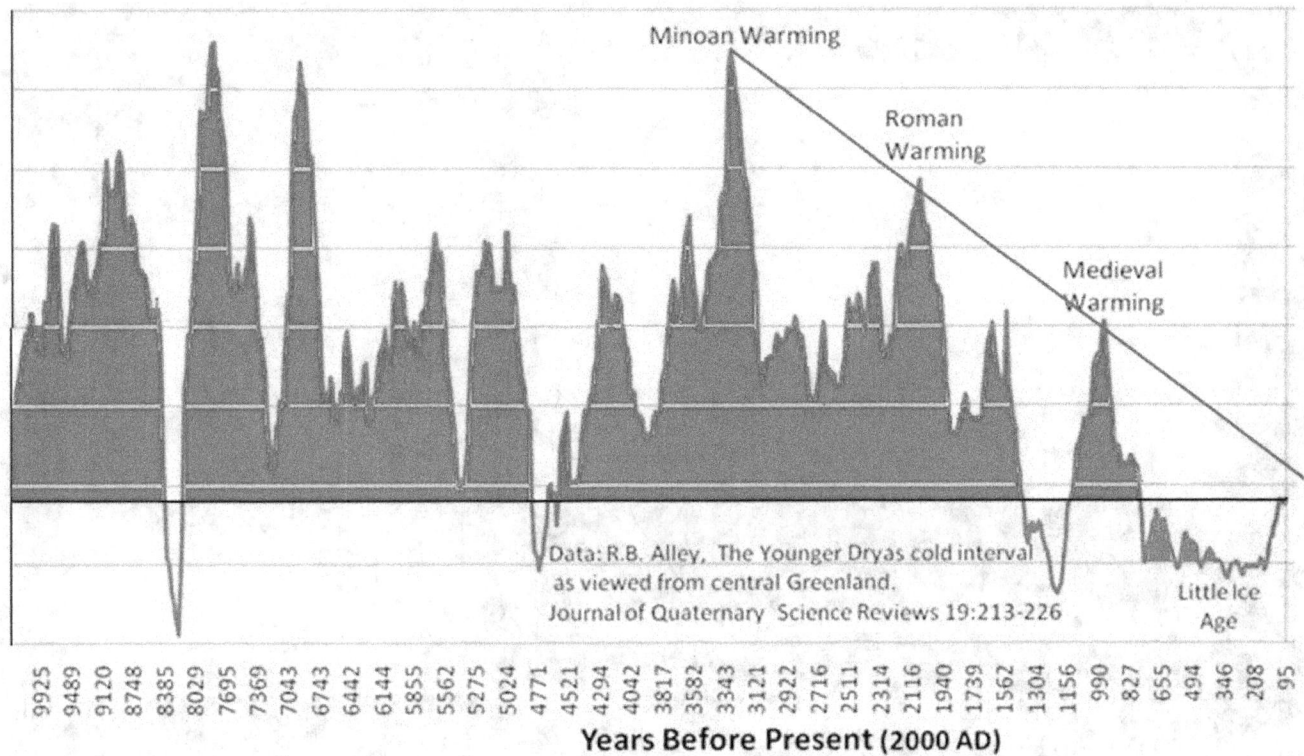

We have seen warm periods and cold periods; and we have seen the warm periods getting progressively weaker, and the cold periods progressively colder. We have also seen large climate recoveries occurring, from the cold periods, such as from the Little Ice Age in the 1600s, all the way to the peak of the modern global warming period in the late 1900s.

Climate is a critical factor for us all

Climate is a critical factor for us all, because it affects agriculture that our food supply depends on. Warm periods are highly productive periods for agriculture.

The Little Ice Age a period of mass starvation

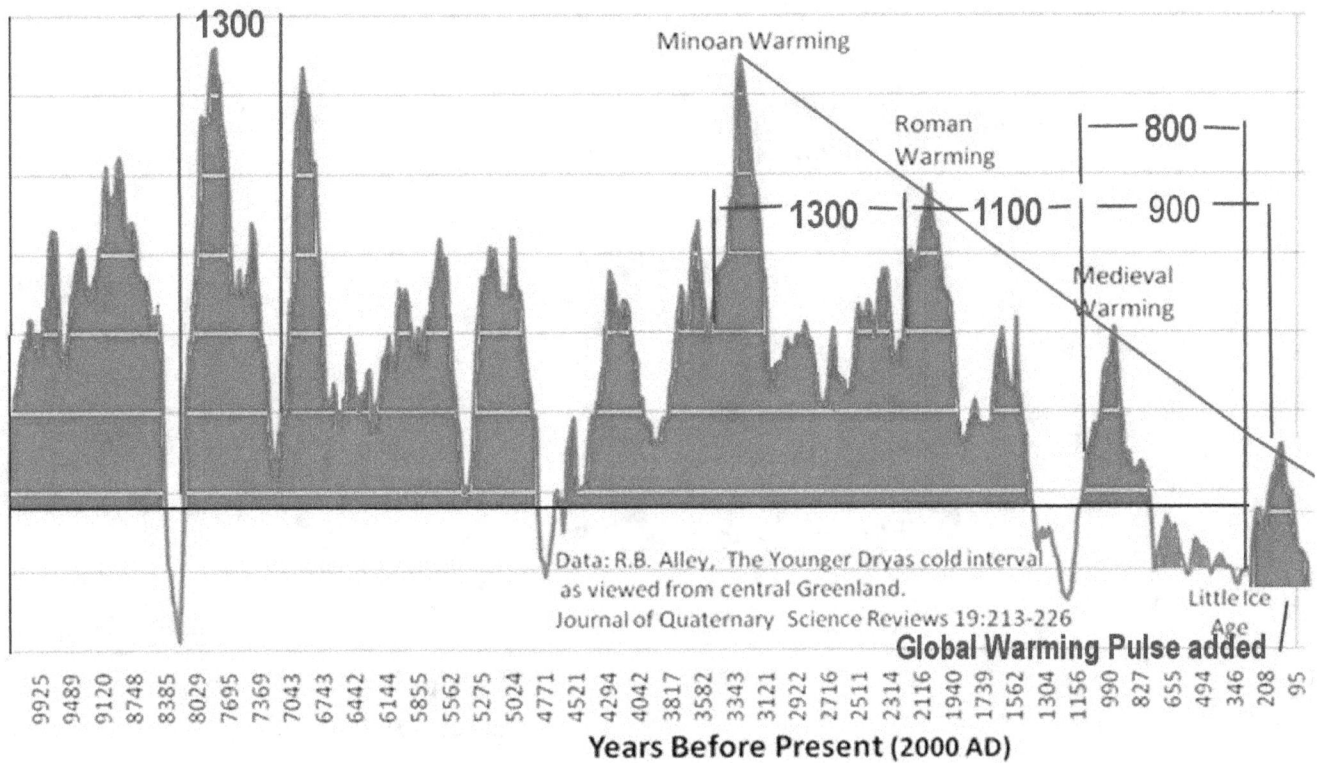

Unfortunately we are already past the peak of the great global warming that had rescued us from the Little Ice Age. The Little Ice Age had been a period of depressed agriculture and consequent mass starvation.

We are now in a period of global cooling

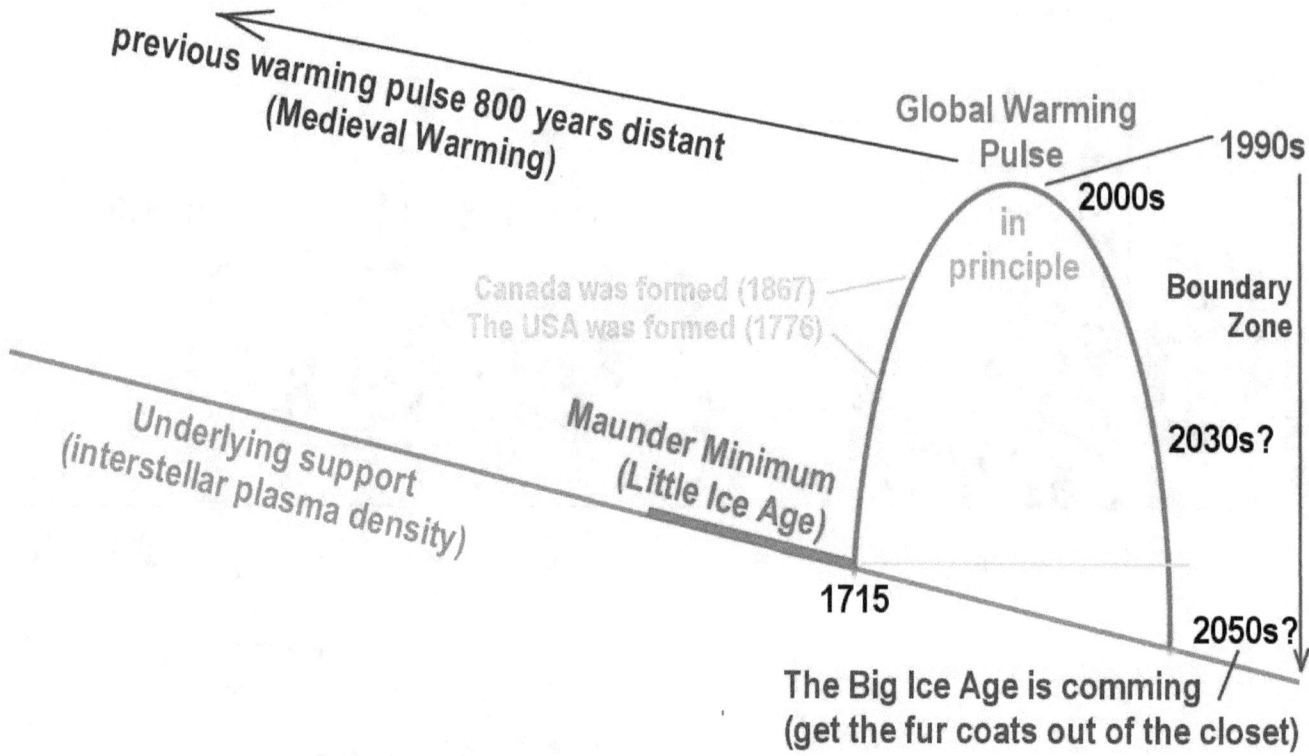

We are now in a period of global cooling again. We see evidence of another climate collapse being in progress towards another Little Ice Age in the near term that promises to collapse into the long-awaited Big Ice Age, except this time with a 7-times larger world population than we had in the Little Ice Age.

Crop yields are diminishing

The climate collapse is happening while food production is already stressed to the limit. Crop yields are diminishing under conditions of drought and increasing climate fluctuations that result in evermore crop failures.

World grain production has already dropped

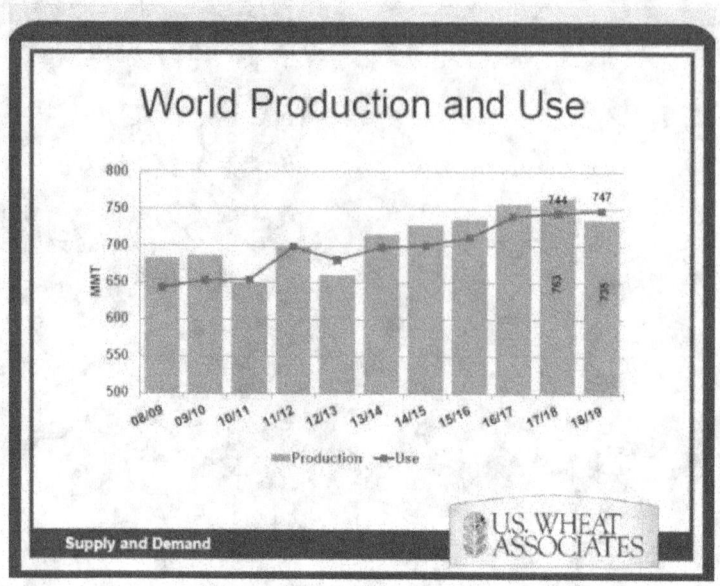

While we don't face a food-crisis yet, much less a political crisis over food shortages, the world grain production has already dropped below requirements. The difference will likely be supplied from the carry-over stocks. The point is, that the ongoing climate collapse is already being felt in agriculture.

The entire world is affected by the changing global climate

NASA - Earth from Apollo 16
wikipedia

Because the entire world is affected by the changing global climate, and with it the fate of entire nations hangs in the balance, it becomes imperative for society to explore the principles that cause the changes, and their increasing expressions in the future, in order that compensating infrastructures can be created with which to assure the continuing food supply for humanity in the rapidly changing world.

Food and climate are presently linked

Food and climate are presently linked. We need to un-link them. For this very-large-scale infrastructures are required, such as indoors agriculture on a global scale.

You may say that this is impossible. It is too big. It cannot be done. There isn't enough money in the world to do it. The reality is, that the challenge appears huge in our eyes, only because we think in small terms. In galactic terms the Ice Age climate change is not extraordinary at all.

This means, that our becoming vulnerable to the Ice Age climate change, in the period ahead, is merely a reflection of us thinking in small terms. Our thinking is not yet brought into the context with the dynamics of the Universe. We expect the Universe to adjust itself to our small-scale thinking. How stupid of us! If we have so little love for our humanity that we find it too expensive to protect our food supply, then we will die. We expect the Universe to change, so that we won't have to bother to live like human beings, and use the resources we have at hand to build ourselves the needed infrastructures for our continued living in an Ice Age world. Shouldn't we rather adjust ourselves to the dimensions of the Universe, and meet its challenge. The universe is challenging us to grow up; to develop truthful science; to up-lift economics; and to up-lift our perception of ourselves.

The large-scale infrastructures can be build that secure our future

So it is that before the large-scale infrastructures can be build that secure our future existence, extensive scientific explorations are required for us to come to terms with the cosmic dynamics that affect our Earth, our climate, and with it our food supply. Without the science being honestly developed, the imperative cannot be recognized for what needs to created for humanity to secure its future, versus having none.

Barriers set up against science, to keep it small

NASA - Earth from Apollo 16
wikipedia

The science-issue that is involved, is that big. But here we face some significant barriers, intentionally created barriers, barriers set up against science, to keep it small - and those appear to be monumental barriers.

These barriers are regarded by the elite like holy cows and are vehemently defended by them, and society has been carefully guided to fall into this trap. Nevertheless, with the fate of many nations now hanging in the balance in the near-term already, it has become imperative that the barriers be overcome, and they can be overcome by humanity becoming alive again as human beings, who have a place for the truth in the heart.

Breaking through long-standing barriers

Breaking through long-standing barriers cleverly erected

Breaking through long-standing barriers cleverly erected.

H. G. Wells wrote a novel, The Time Machine

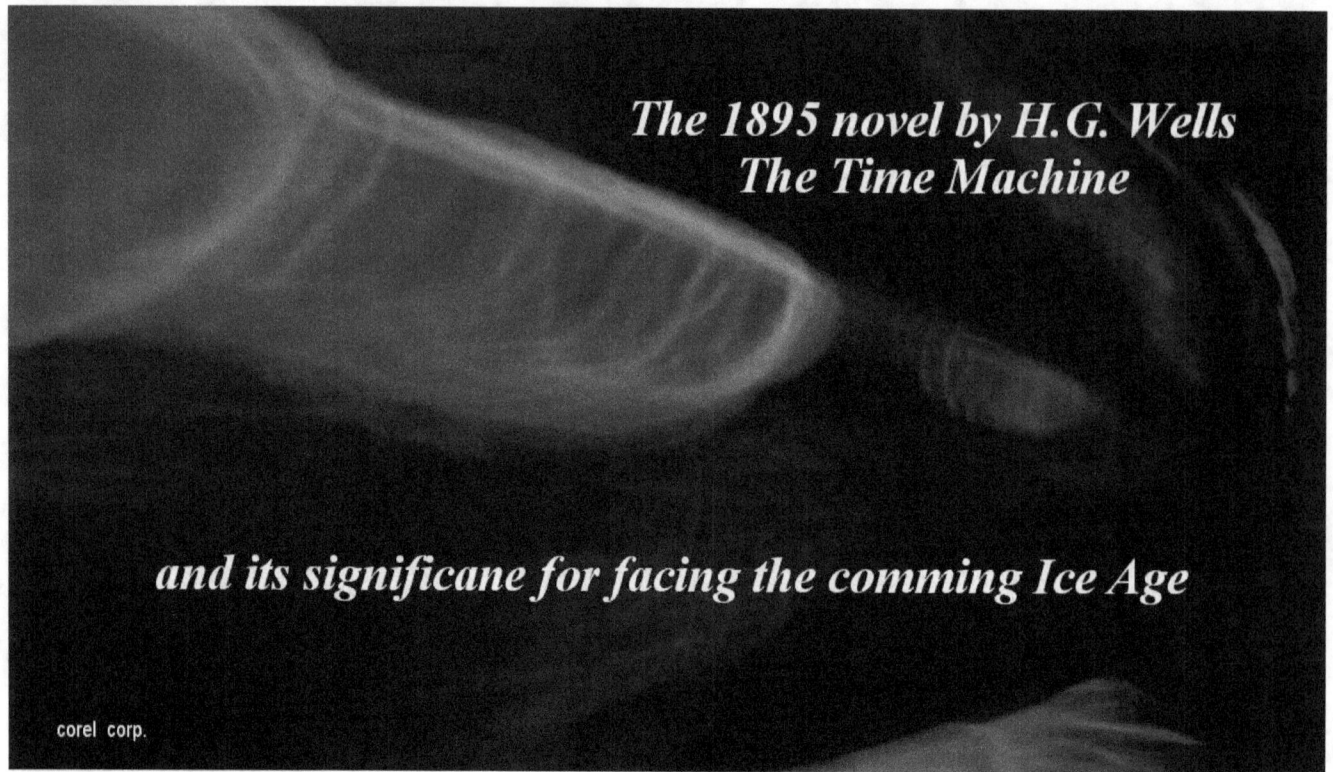

The 1895 novel by H.G. Wells
The Time Machine

and its significane for facing the comming Ice Age

corel corp.

Back in the late 1800s, the British author H. G. Wells wrote a novel, The Time Machine, that apparently changed the world. In the novel the inventor of a time machine travels far into the future where he encounters world of docile beautiful people, the Eloy, living without toil, without a care, abundant with food. When his time machine was stolen, he searches for it and discovers still another type of people, a dirty science and machine-loving people, the Morloch, who live underground, who keep everything functioning. He discovers from them the secret behind the beautiful docile people, the Eloy. The Morloch maintain them as livestock and eat them for breakfast.

H. G. Wells was saying to the elite

Herbert George Wells in 1943 the man who changed the course of science

H. G. Wells was saying with his novel to the elite of his day, you must not allow science to develop in society, or else society will eat you for breakfast. It appears the message was gradually understood. It has been reported that in the 1920 a debate was going on among the elite as to what to do with science. While science cannot really be stopped, it can be controlled and perverted.

It was in the 1920s, when the orbital cycles theory was invented

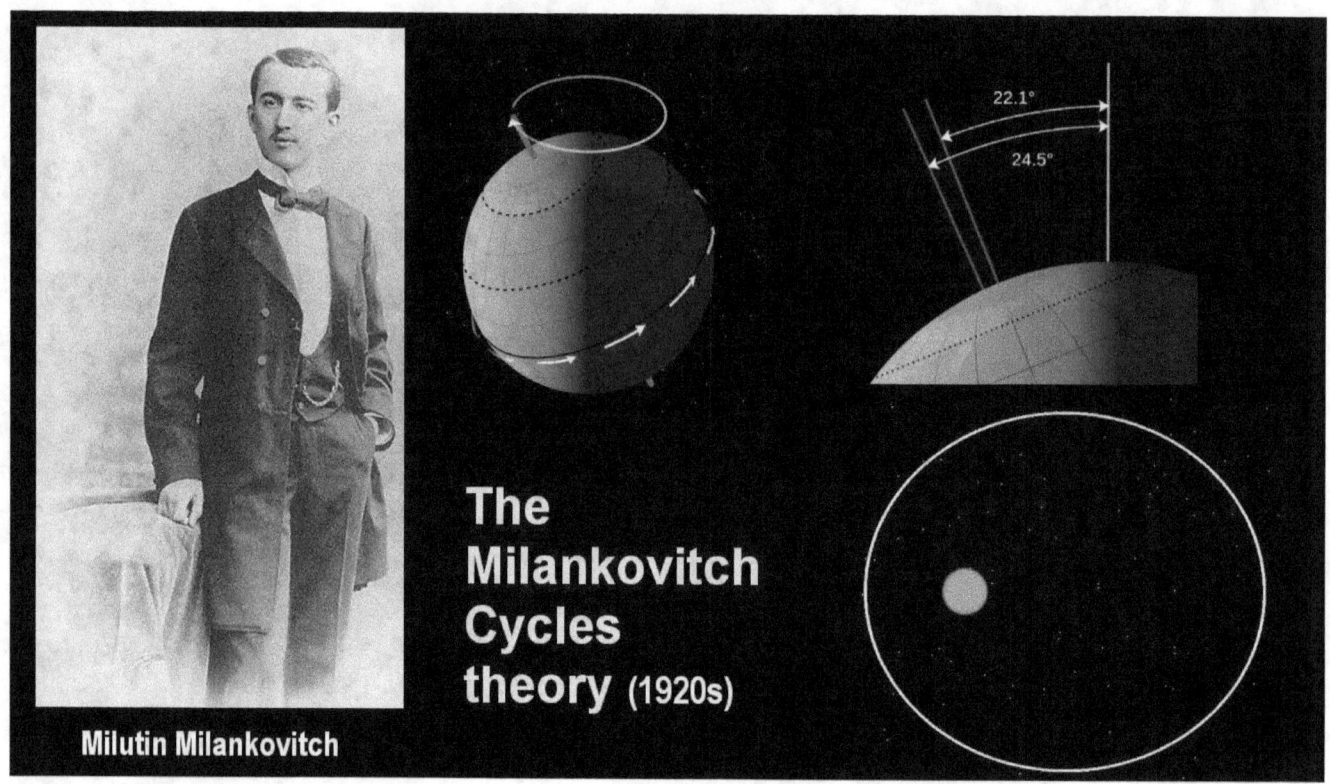

It is interesting to note that it was in the 1920s, when the orbital cycles theory for the ice ages was invented. The theory is based on the premise that the Sun is an invariable constant for all climate considerations. It was reasoned from this premise that if the Sun remains constant, then the ice ages must be caused by cyclical variations of the orbit of the Earth around the Sun.

Milankovitch, a mathematician, combined the orbital variations

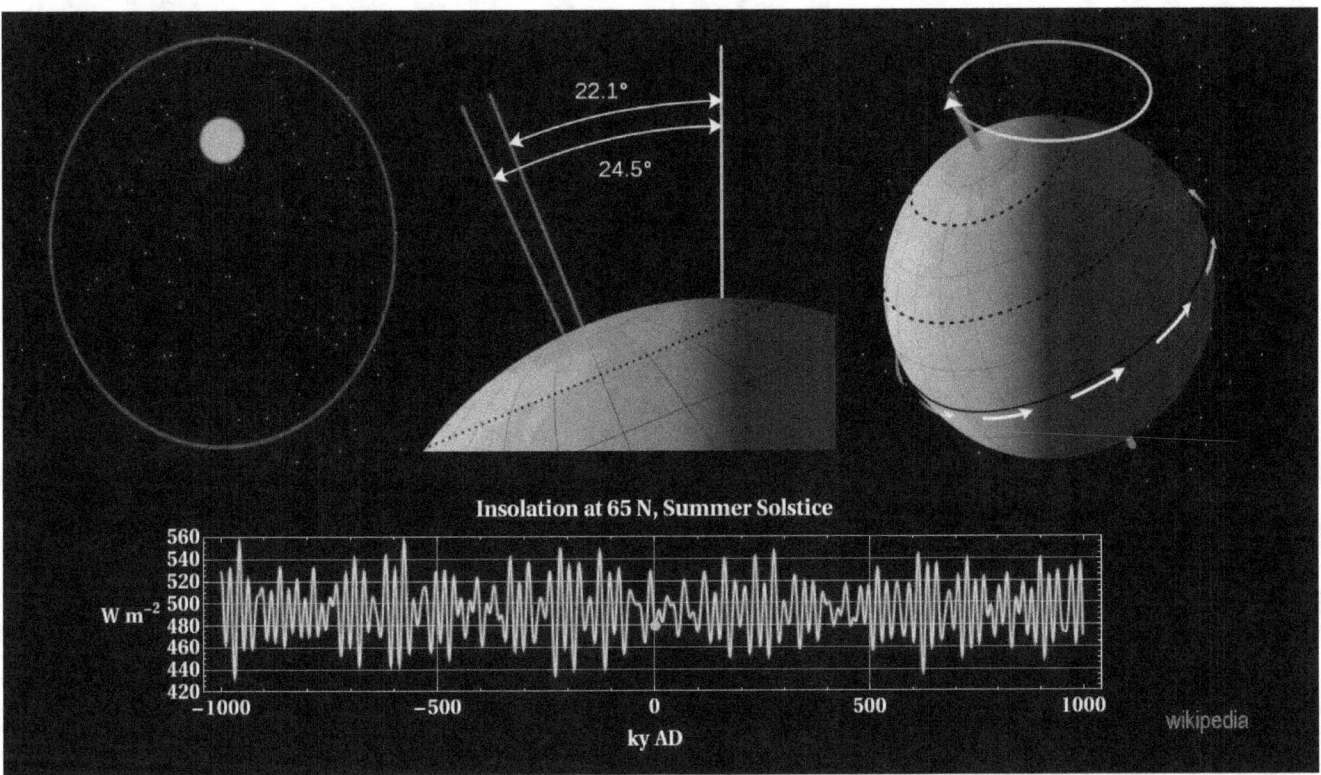

Milankovitch, a mathematician, combined the three known long-term orbital variations, with a 26,000-years cycle time, a 41,000-years cycle time, and a 100,000-years cycle time, and concluded that these cycles, minute as they are, overlap and cause the ice ages, for reasons that they affect the rate of solar irradiation.

The total solar irradiation of the Earth never varies

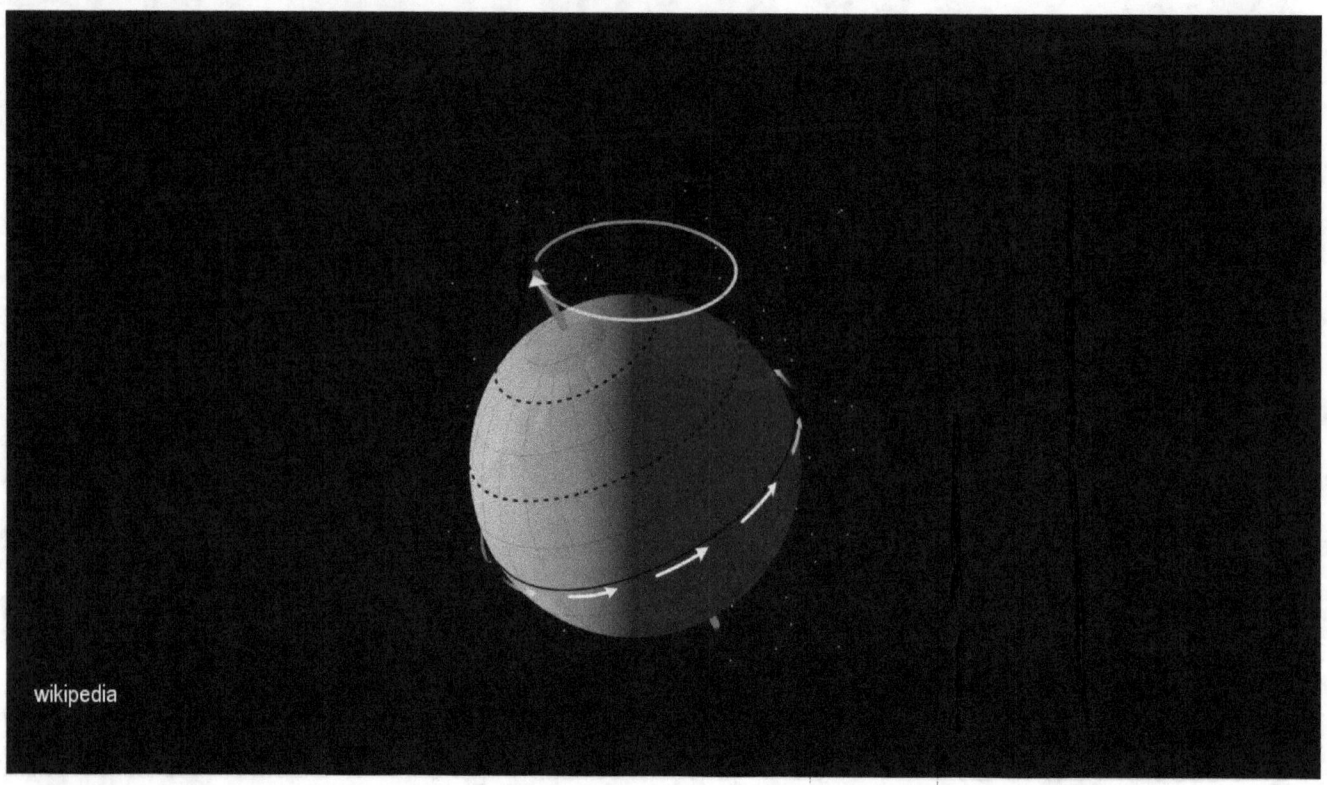

He 'forgot' to mention that the total solar irradiation of the Earth never varies, no matter how its spin-axis is oriented.

Total energy received from the Sun remains the same

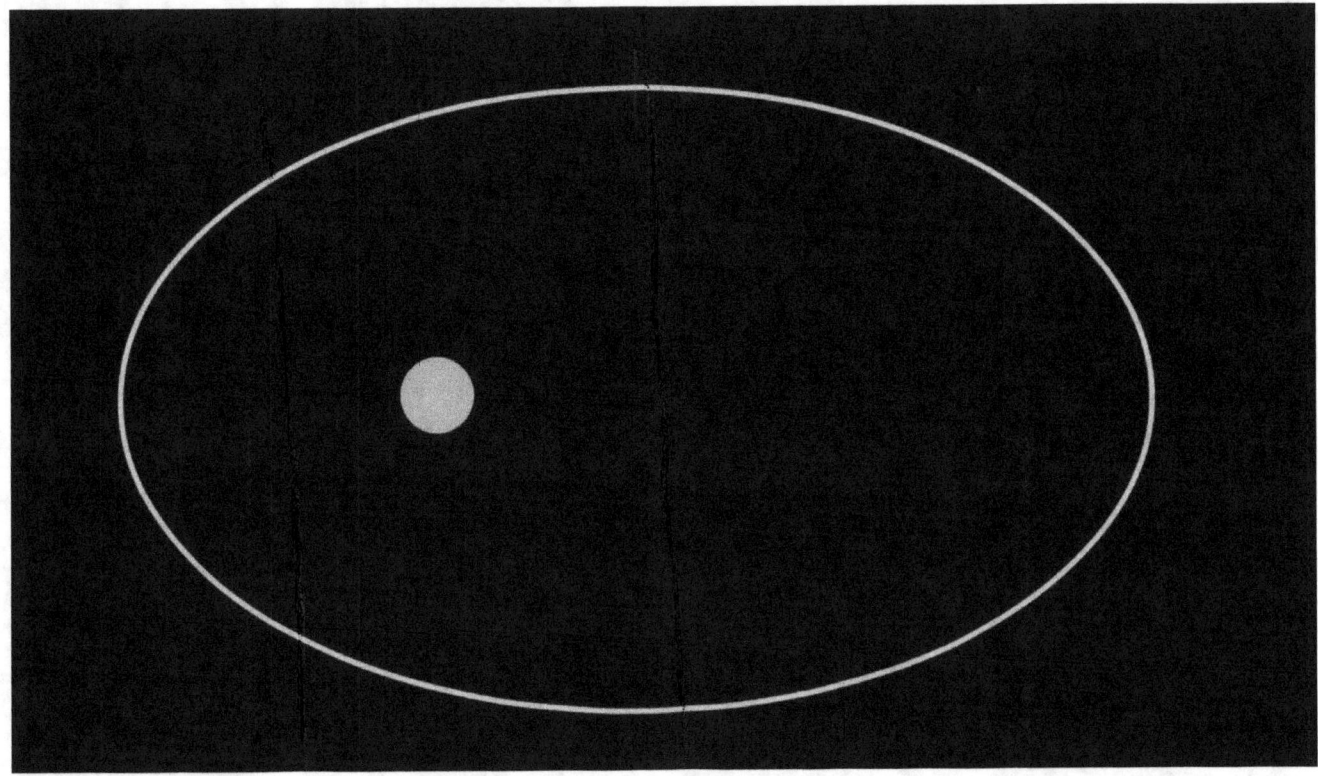

Also, he failed to recognize that the total energy received from the Sun remains the same, regardless of how its orbital eccentricity varies.

Johannes Kepler had made those basic recognitions

Johannes Kepler had made those basic recognitions already back in the 1600s.

Nor did Milankovitch say a word about the sunspot cycles

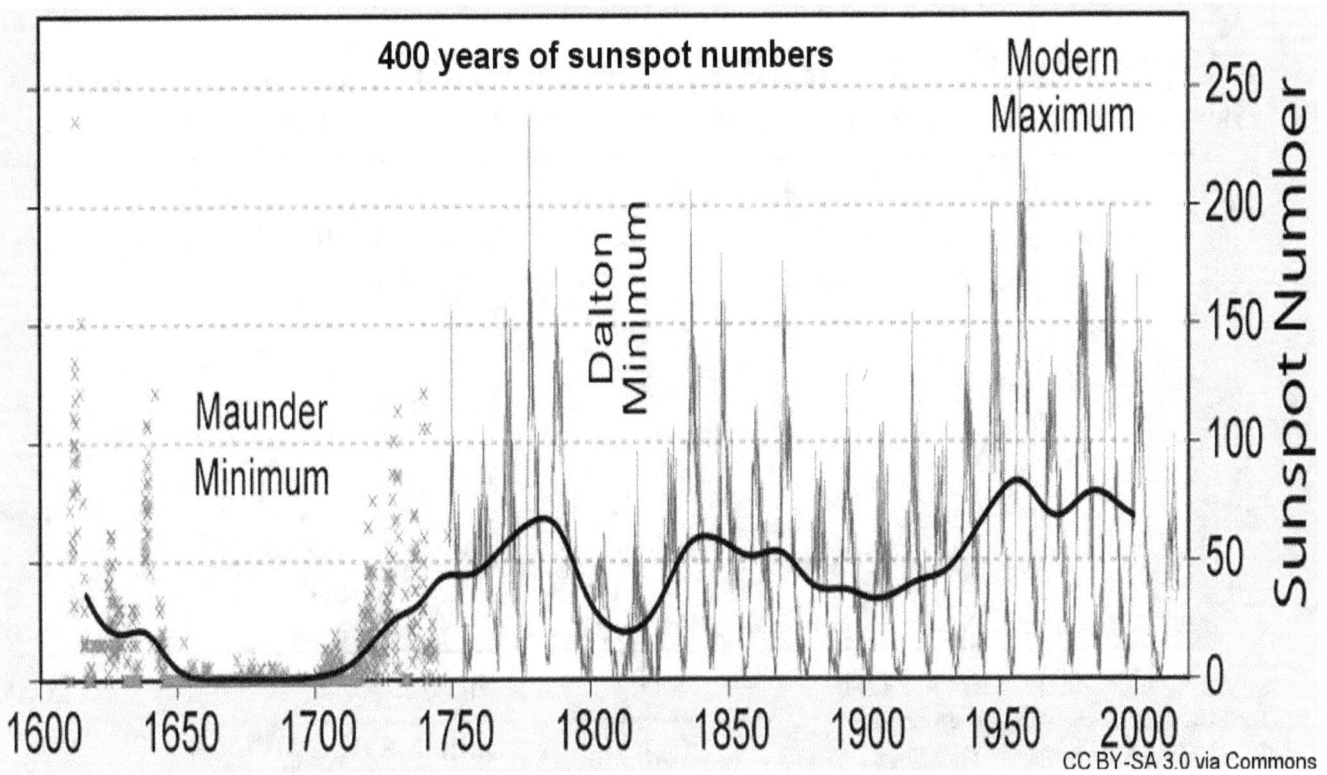

Nor did Milankovitch say a word about the sunspot cycles, which had been carefully recorded from the 1600s onward, which accord closely with changes in historic climate. It had been no secret in the science community, in Milankovich's time, that during the Little Ice Age of the 1600s, when many people were staving to death, no sunspots had been observed on the Sun, and that from the 1700s on, when the sunspots were back, the world became warmer again and the starvation ended.

Some people may have pointed out that the coincidence of the changing sunspot numbers with the changing climate, doesn't really prove anything. And they would have been right, in saying this, because no measurable connection did exist at the time, that links the Sun with the climate on Earth.

Proof that the Sun is the Earth's climate master

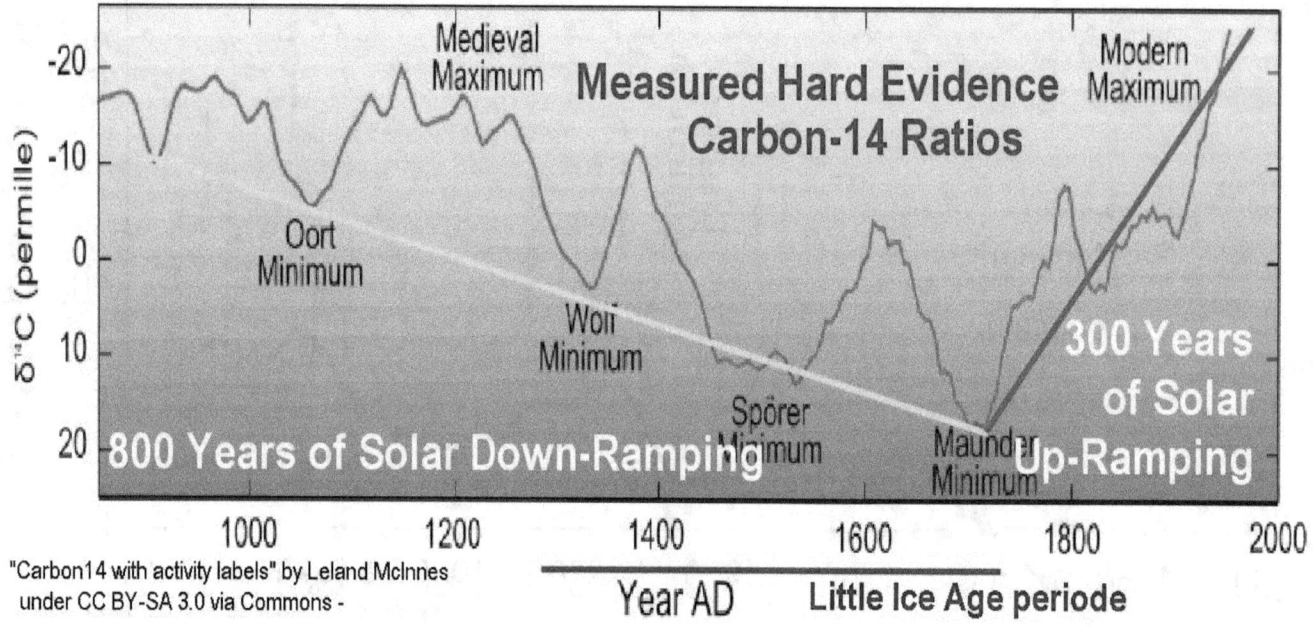

This missing proof was delivered many decades later, when Carbon-14 measurements of historic ice samples indicate with measured clarity, that solar activity really had changed in accord with the historic sunspot numbers, and that the climate on Earth therefore had changed in accord with changes in solar activity. The Carbon-14 measurements thereby certify that the Sun is the Earth's climate master.

Carbon-14 is a radio-isotope that is produced exclusively by cosmic-ray interaction with the atmosphere. The isotope variances reflect variances in solar activity, because in times of weak solar activity, when the plasma corona around the Sun is weak, which traps a portion of solar cosmic-rays, the resulting solar cosmic-ray flux is less impeded, so that high volumes of it affect us on Earth.

The Carbon-14 measurements 'certify'

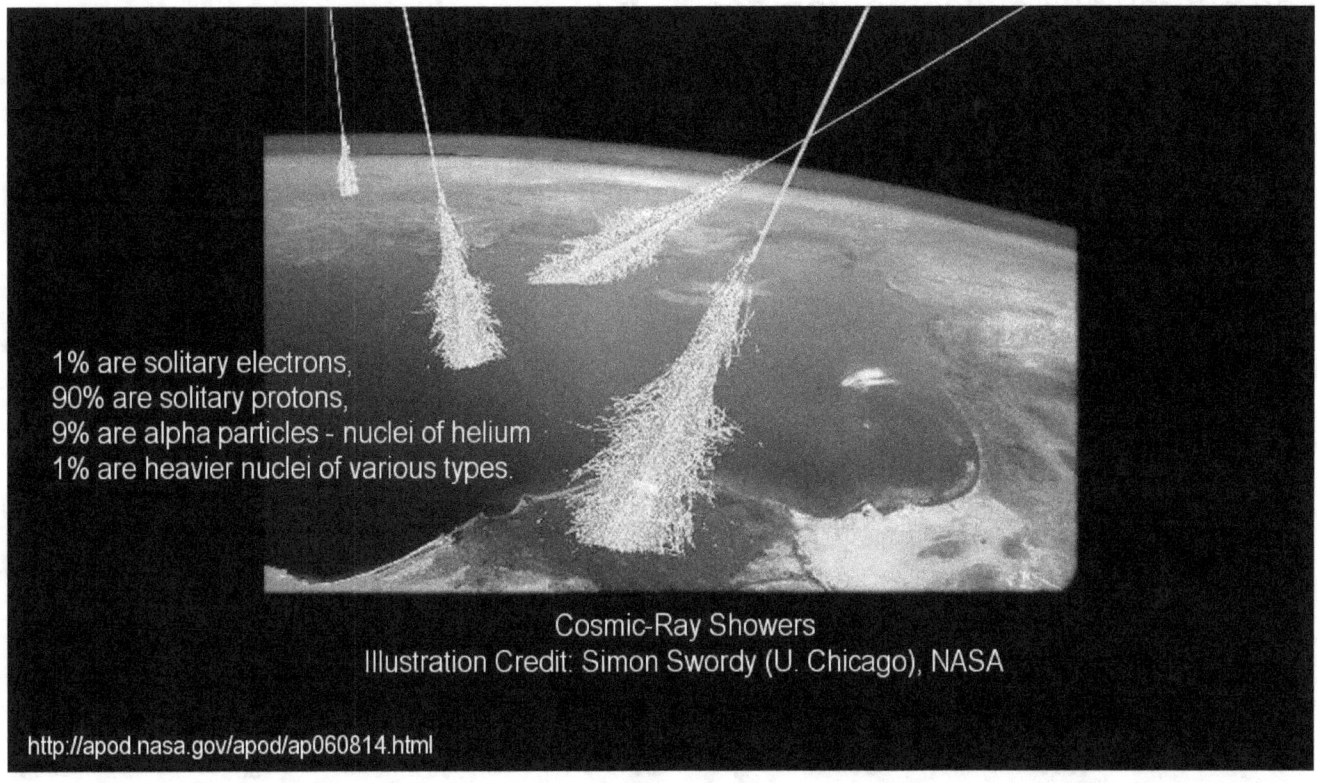

Thus, the Carbon-14 measurements 'certify' conclusively that the Sun is definitely not an invariable climate factor, but that it affects the climate on Earth in a big way, via changes in cosmic-ray flux.

But do the Carbon-14 measurements, by themselves, prove that the cosmic-ray flux did come from the Sun?

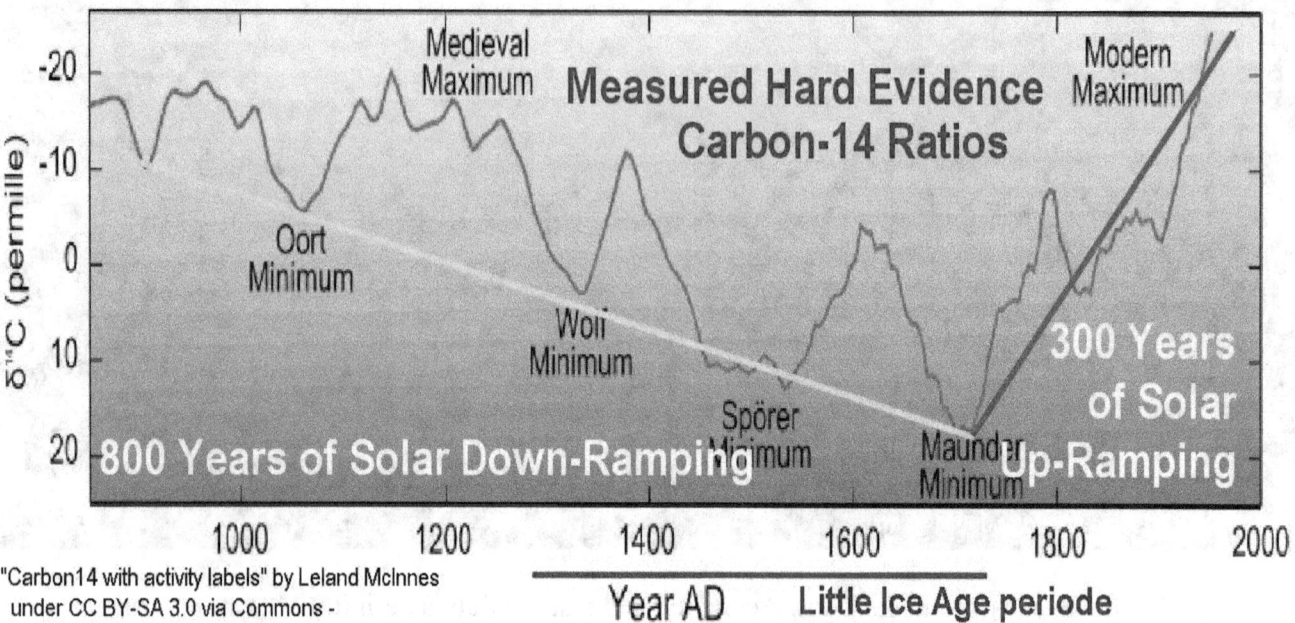

But do the Carbon-14 measurements, by themselves, prove that the cosmic-ray flux that has created the Carbon-14, which is being measured, did come from the Sun?

Can we take these measurements as proof? No, the measurements do not prove that the fluctuations were caused by the Sun.

The needed proof was delivered by Berillium-10

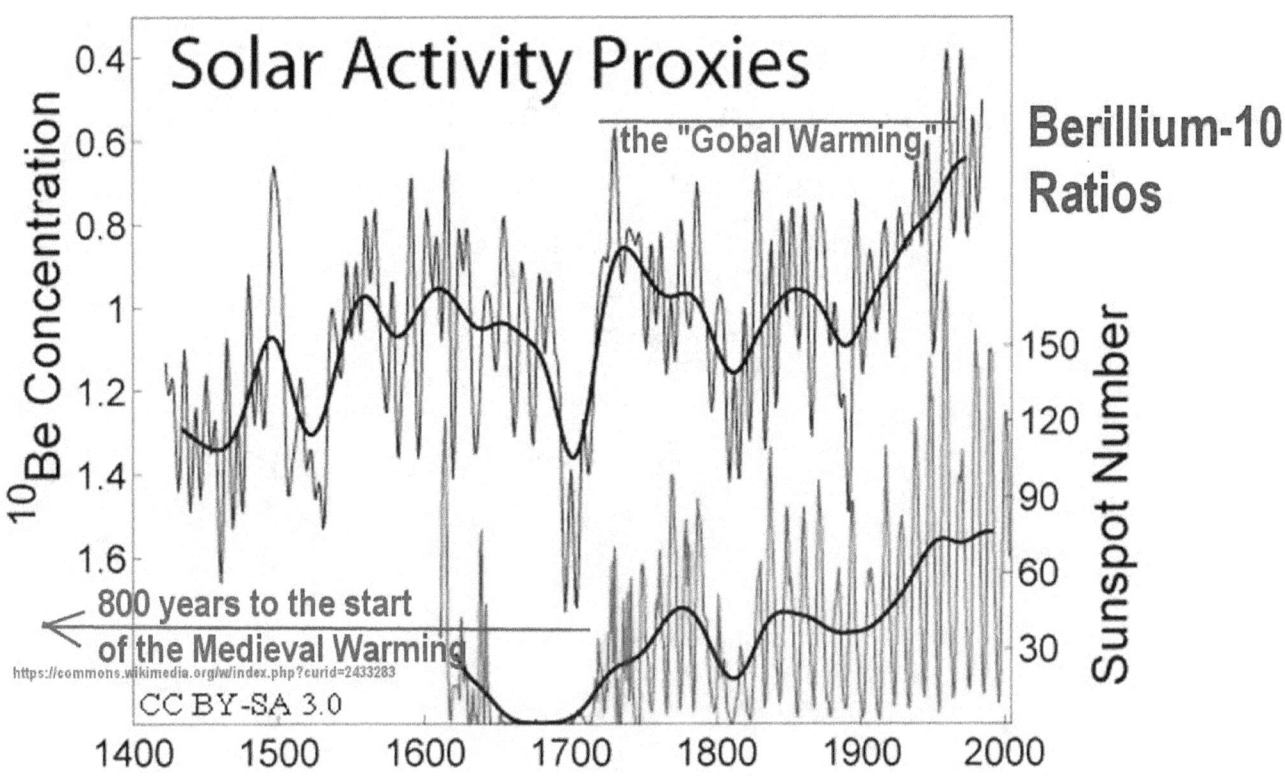

The needed proof was delivered by measurements of the Berillium-10 isotope. Berillium-10 is another radioisotope that is produced in the atmosphere by cosmic-ray interaction.

The Berillium-10 measurements yield higher-resolution results

Reaction products of primary cosmic rays, lifetime and reaction

Tritium (12.3 a): 14N(n, 3H)12C (Spallation)
Beryllium-7 (53.3 d)
Beryllium-10 (1.6E6 a): 14N(n,p a)10Be (Spallation)
Carbon-14 (5730 a): 14N(n, p)14C (Neutron activation)
Sodium-22 (2.6 a)
Sodium-24 (15 h)
Magnesium-28 (20.9 h)
Silicon-31 (2.6 h)
Silicon-32 (101 a)
Phosphorus-32 (14.3 d)
Sulfur-35 (87.5 d)
Sulfur-38 (2.8 h)
Chlorine-34 m (32 min)
Chlorine-36 (3E5 a)
Chlorine-38 (37.2 min)
Chlorine-39 (56 min)
Argon-39 (269 a)
Krypton-85 (10.7 a)

wikipedia - Cosmic Rays

Berillium-10 is produced in the atmosphere in a different manner, by spallation, instead of by neutron activation as Carbon-14 is produced. The Berillium-10 measurements thereby yield higher-resolution results

The berillium-10 ratios follow the solar cycles closely

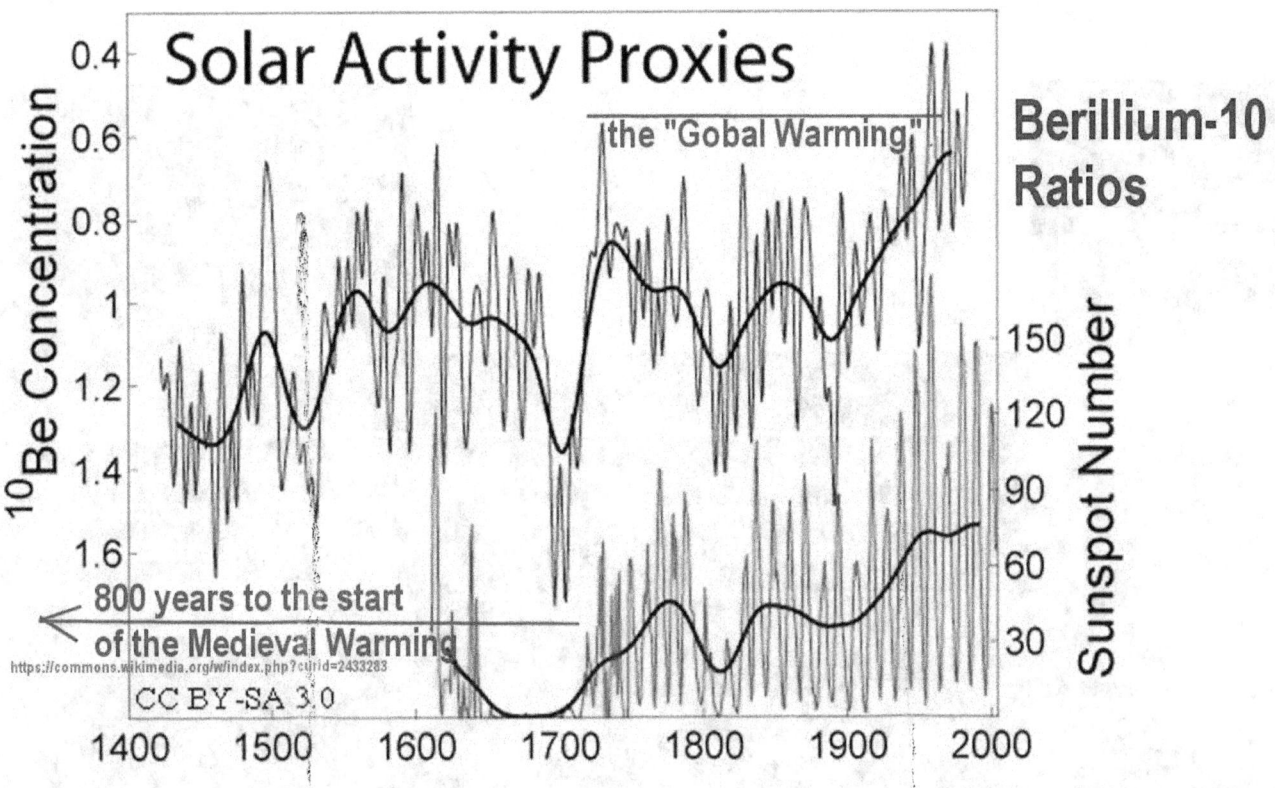

And what are the results that we see? We see that the measured berillium-10 ratios follow the solar cycles closely, as we have them recorded in sunspot numbers. This coincidence certifies the existence of solar cosmic-ray flux, and it certifies that solar cosmic-ray flux is affecting the Earth.

This means that the three measurements combined, which all tell us the same story, which in combination certify that the Sun is not a constant factor; that it is dramatically changing; that the changing solar activity is affecting the Earth; and that the Sun affects the climate on Earth by changing solar cosmic-ray flux.

But do these measurements proof that changing solar activity is causing the ice ages? No

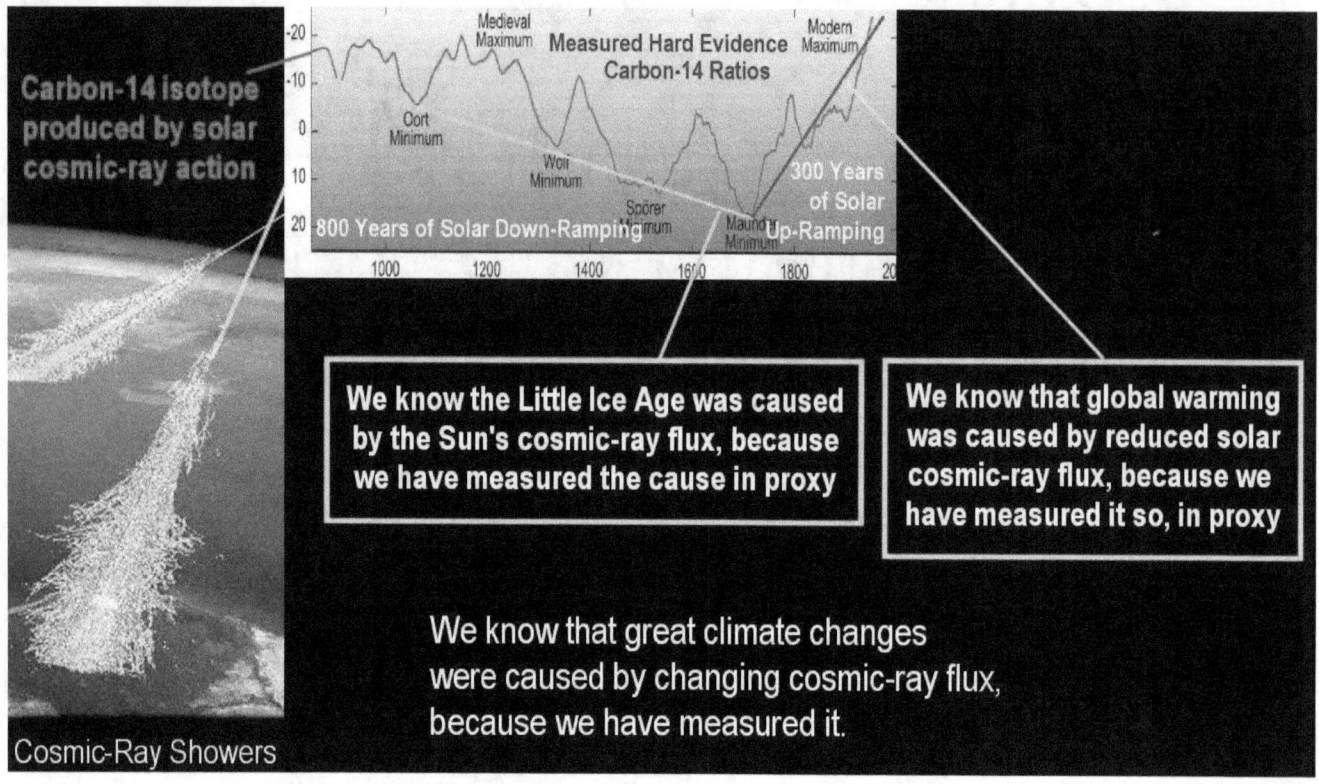

But do these measurements proof that changing solar activity is causing the ice ages? No, they don't prove that.

The measurements only prove that a connection exists between diminishing solar activity and diminishing climate on Earth.

To look deep into interglacial climate history

Only the ice core temperature measurements reach that far with historic temperature measurements, in Oxygen-18 isotope ratios. The historic long-term measurements are produced by the giant ice coring projects that enable us to look deep into interglacial climate history.

The weakening of the warm climate that has been measured

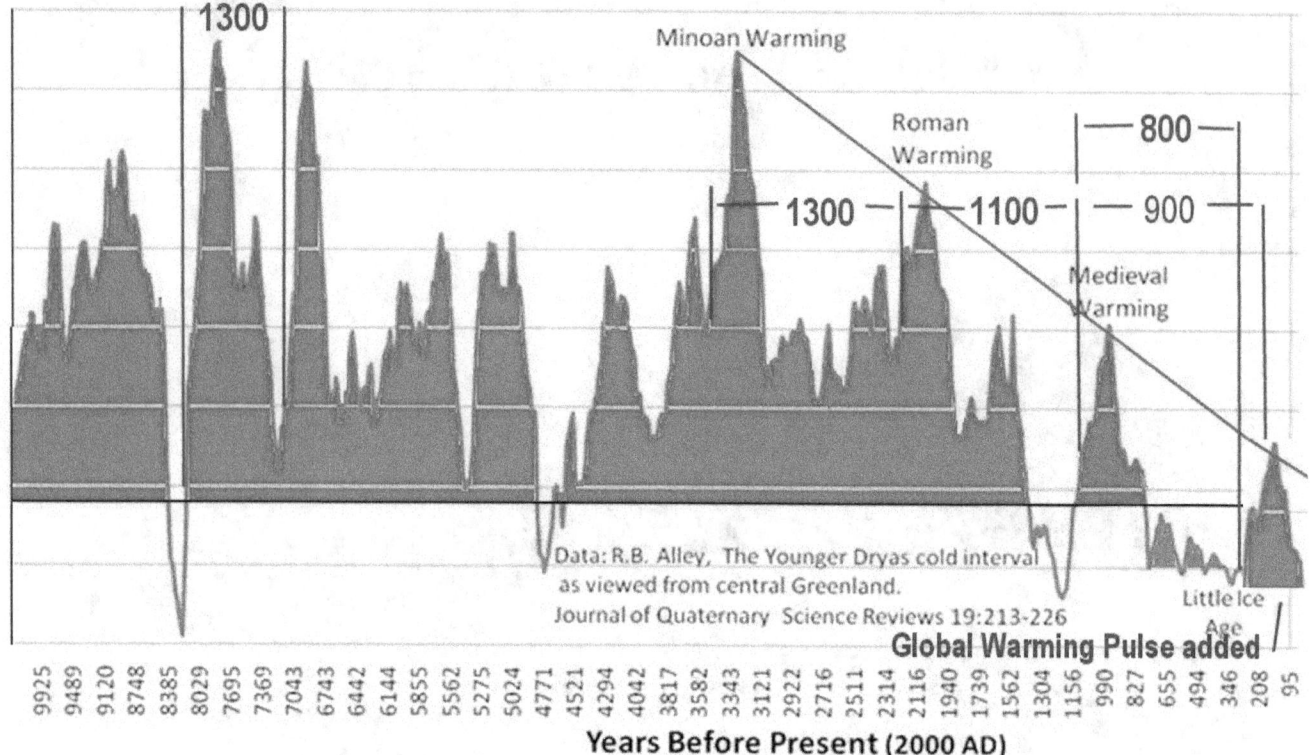

However, the resulting ice core measurements tell us only that our nicely warm interglacial climate is rapidly diminishing. They do not tell us what causes the warm climate to diminish.

While the carbon and beryllium isotope measurements, in turn, tell us with measured certainty, that the weakening of the warm climate that has been measured in the ice core samples, was the result of changing solar activity levels and of correspondingly changing solar cosmic-ray flux, the radio-isotope measurements cannot tell us why the solar activity is diminishing.

Certainty that cosmic factors affect our climate

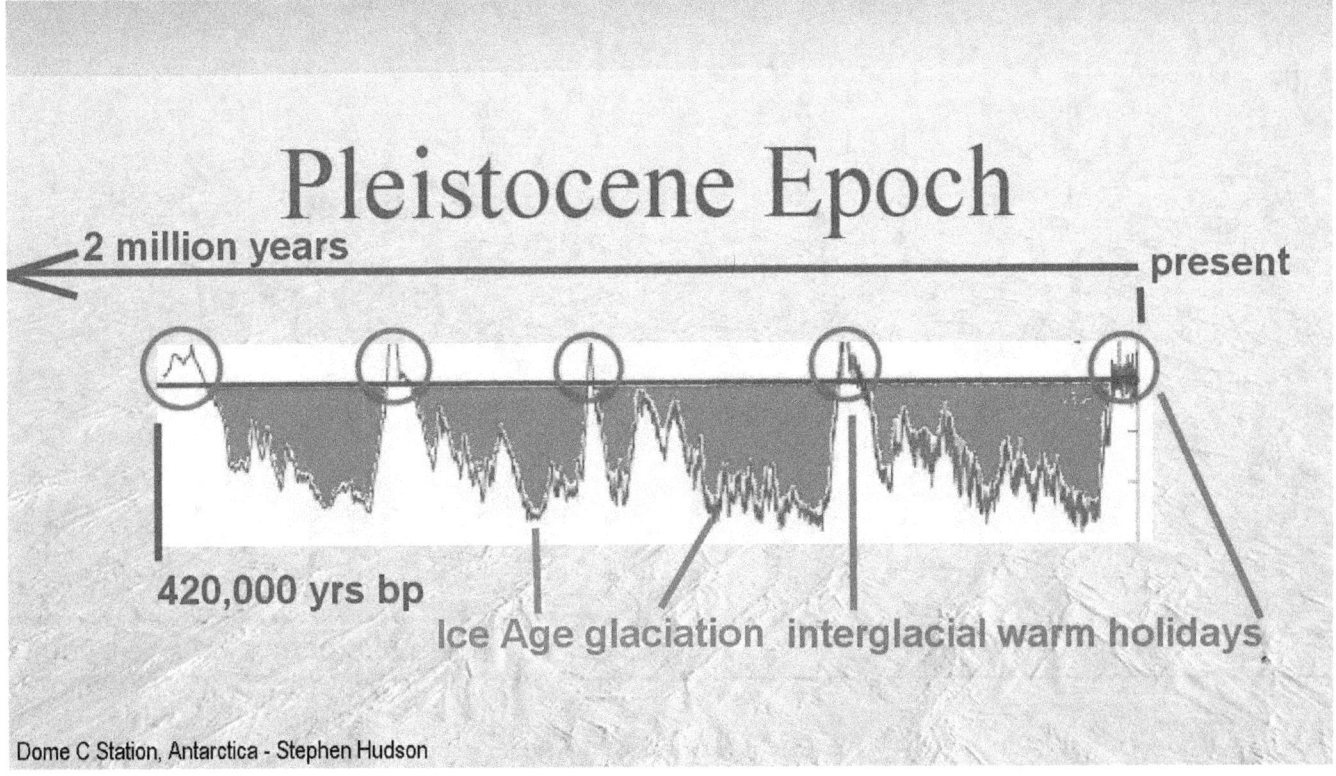

Thus, the isotope measurements cannot tell us either, what causes the dramatic Ice Age climate collapse between the brief interglacial periods that the deep ice core projects in Antarctica tell us of.

All that we get out of these measurements with certainty, at this stage, is that we have measured proof developed, that cosmic factors related to the Sun, affect our climate, and that these can have huge effects, beyond anything we have ever experienced in recorded history.

Ice Ages do not happen gradually

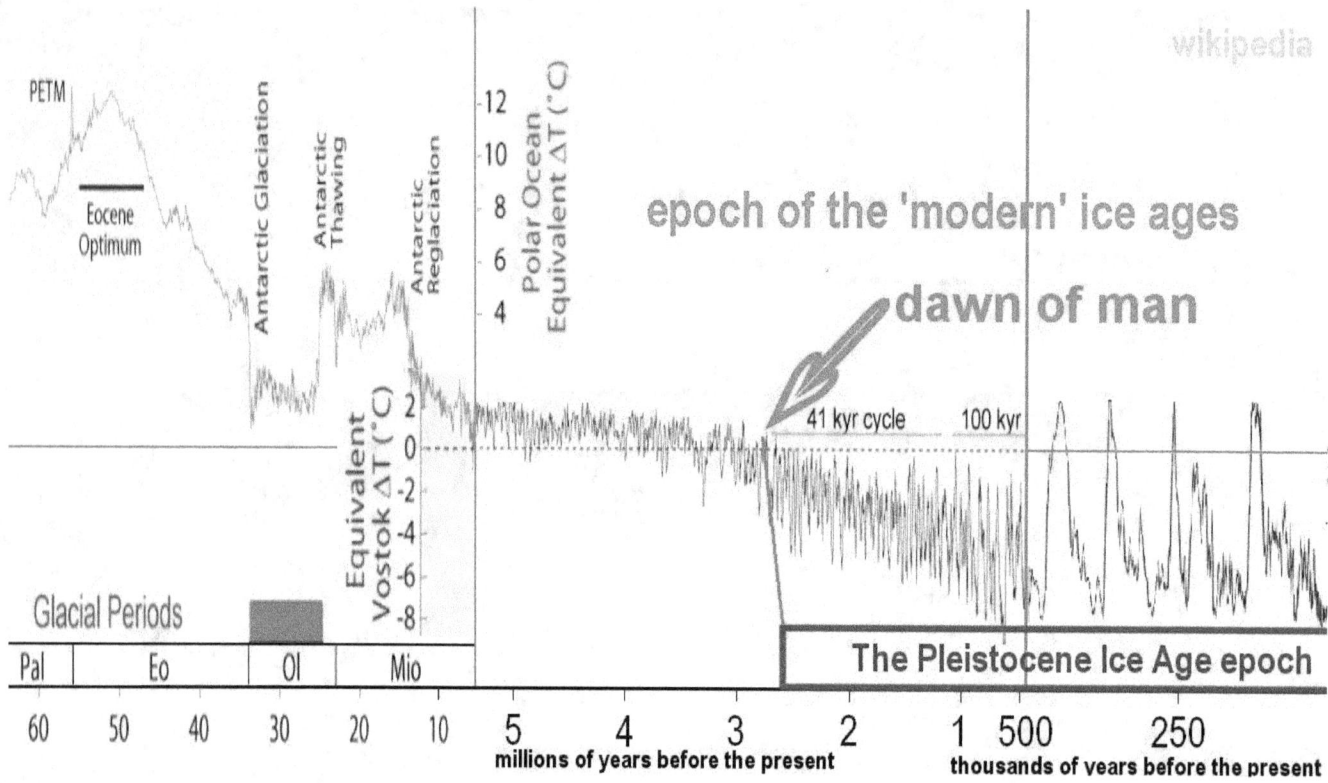

We know from deep-sea sediments, which go far back in time, that Ice Ages do not happen gradually. This is evident in the expanded view on the right.

Ice Ages result from dramatic oscillations of the climate

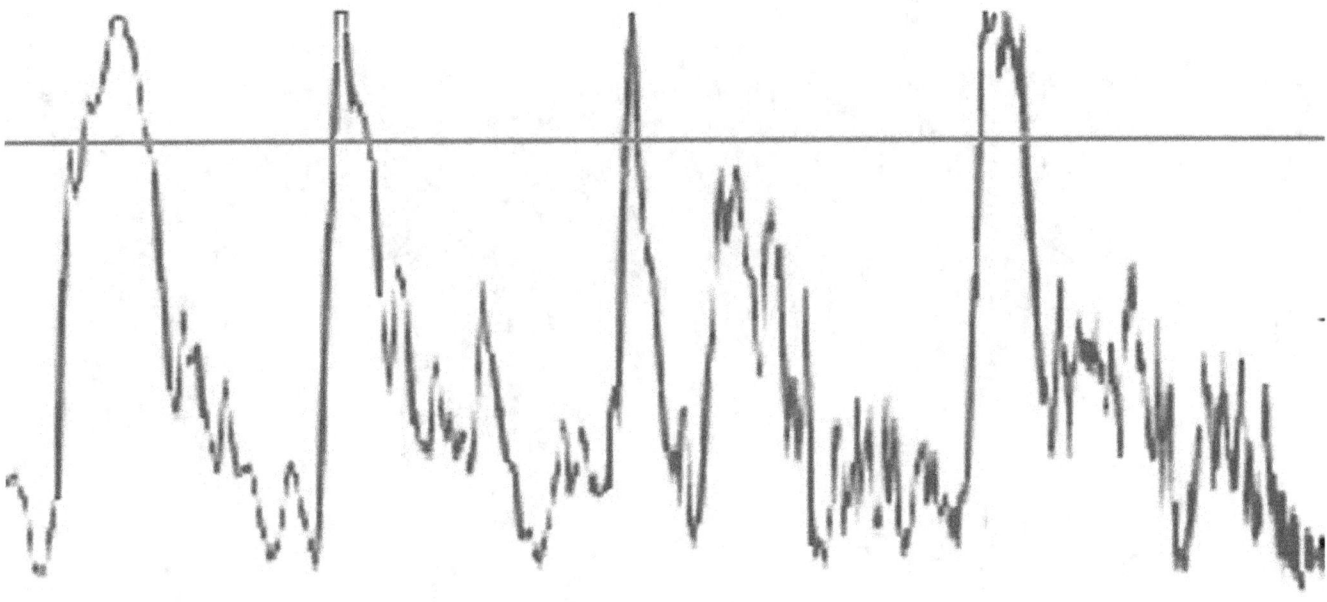

The deep-sea sediments tell us that Ice Ages result from dramatic oscillations of the climate on Earth, with large transitions occurring between the extremes. These oscillations are far larger than what cosmic-ray variations can produce.

Glaciation, like a person falling off a cliff,

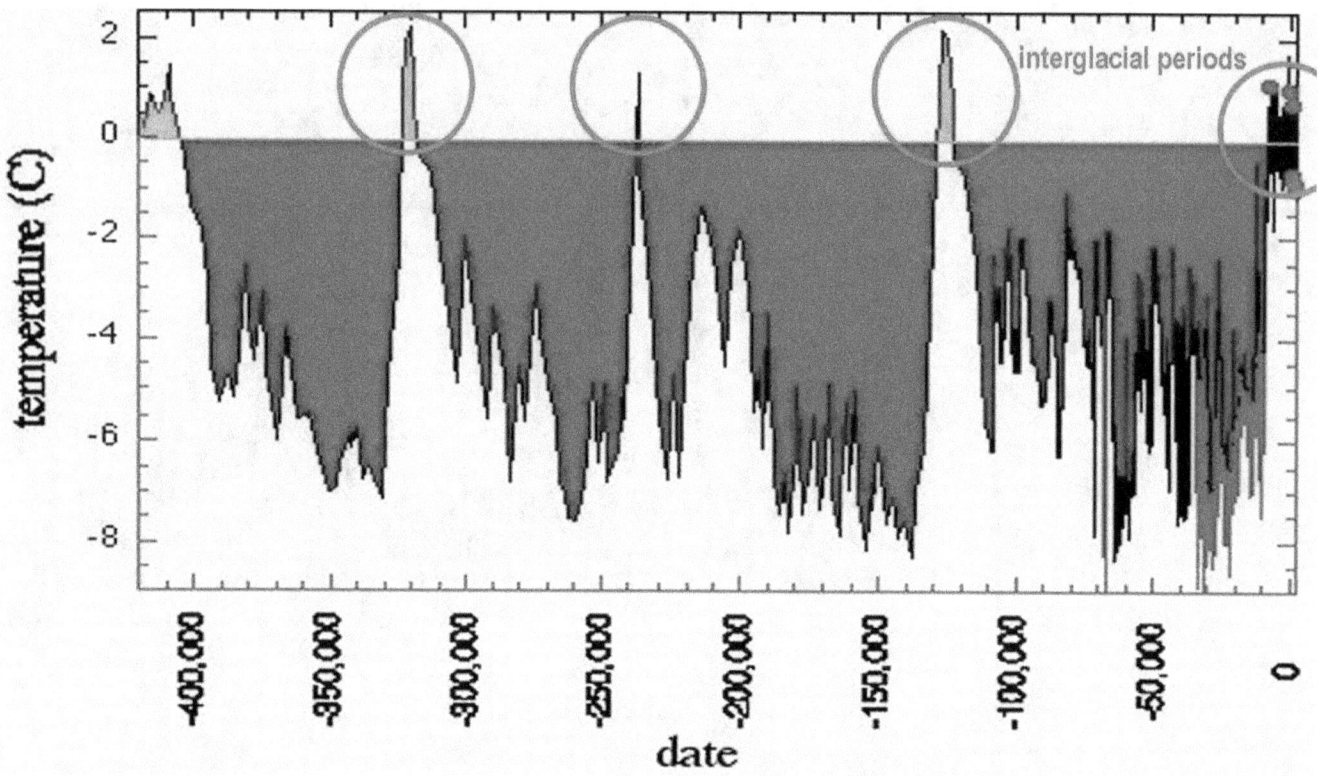

We see the same in ice core samples. We see that when the glaciation climate erupts, it erupts rapidly, like a person falling off a cliff, which is never a gradual affair.

Here too, we see that the glaciation climate takes us deep into unknown territory where unconventional factors evidently come into play.

Cooling had been up to 40 times deeper than during the Little Ice Age

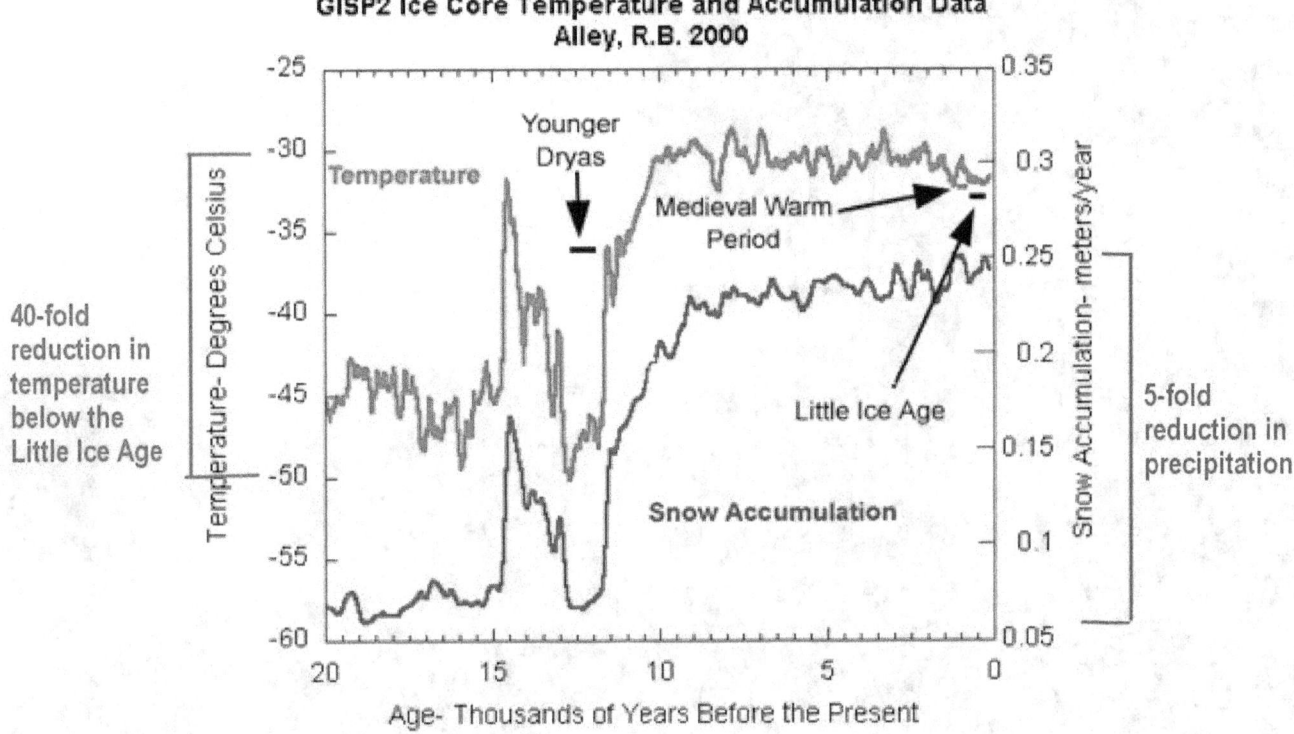

Ice core samples tell us that the big Ice Age cooling had been up to 40 times deeper than the cooling that had been experienced during the Little Ice Age in the 1600s.

While the Little Ice Age cooling had been extremely harsh in consequences, it was minuscule in the overall landscape. In considering that the tiny bit of cooling of the Little Ice Age had wiped out large parts of the populations in some places in Europe, by starvation, we get a faint glimpse of what we are up against when the full Ice Age glaciation begins.

The comparison gives us a sense of the scale of the climate change

The comparison gives us a sense of the scale of the climate change that we are facing in the near term when the Ice Age phase shift happens.

That 80% less precipitation occurs during glacial times

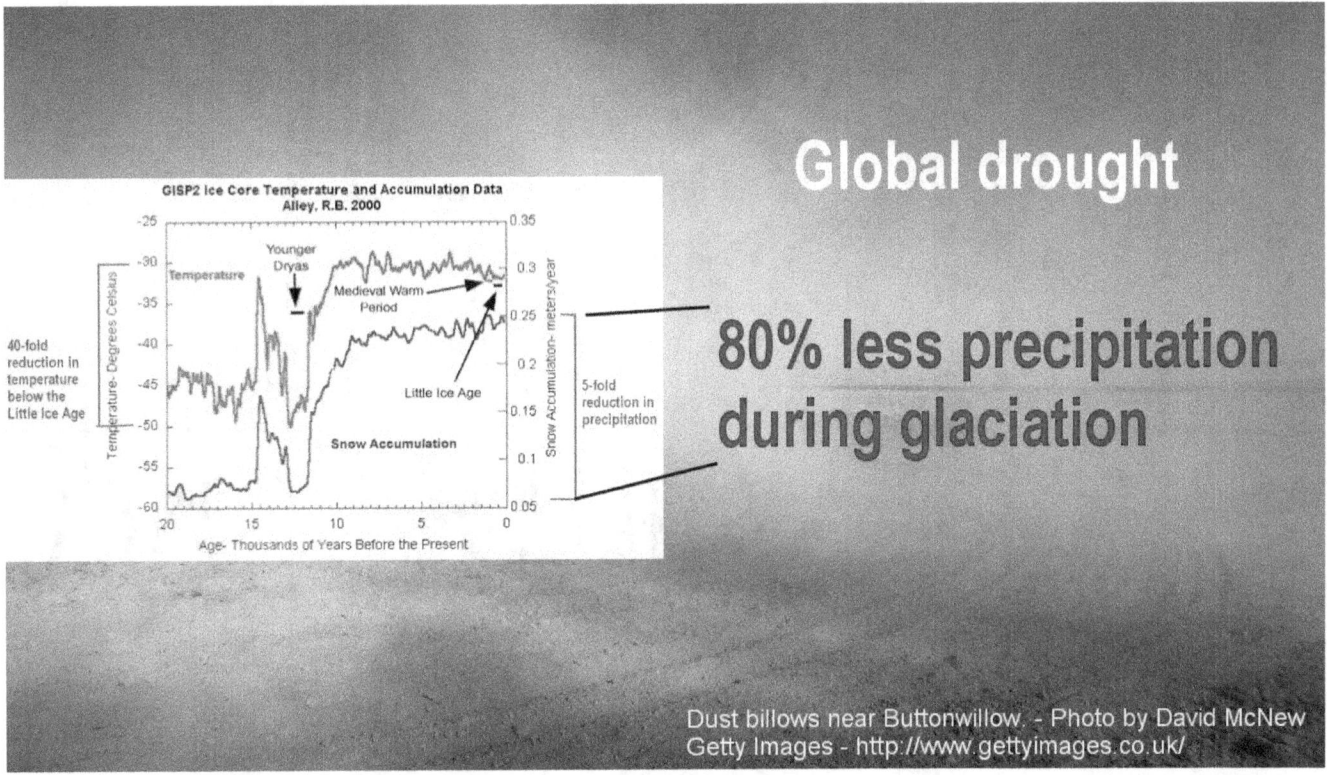

The ice core records also tell us that 80% less precipitation occurs during glacial times. This radical transformation, too, is what we need to develop compensating solutions for.

Ice ages are essentially digital in nature

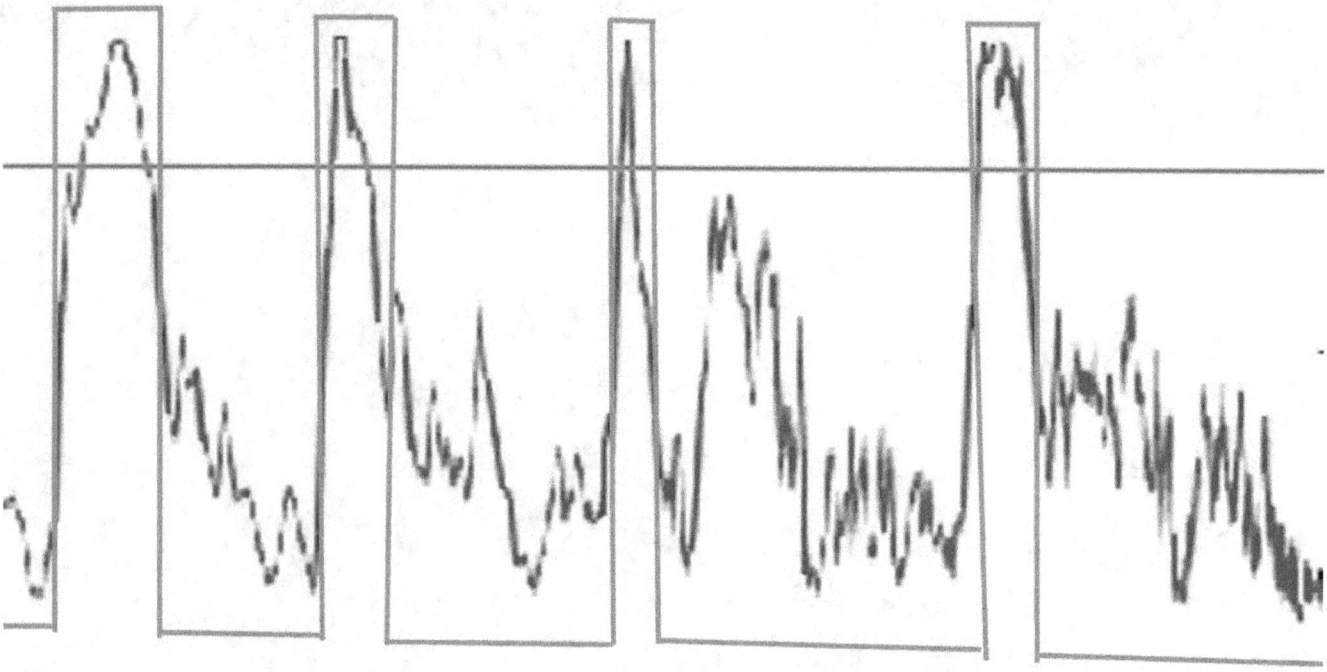

The measurements from deep sea fossils tell us essentially the same story. They tell us that ice ages are essentially digital in nature.

Hope and pray for a soft landing

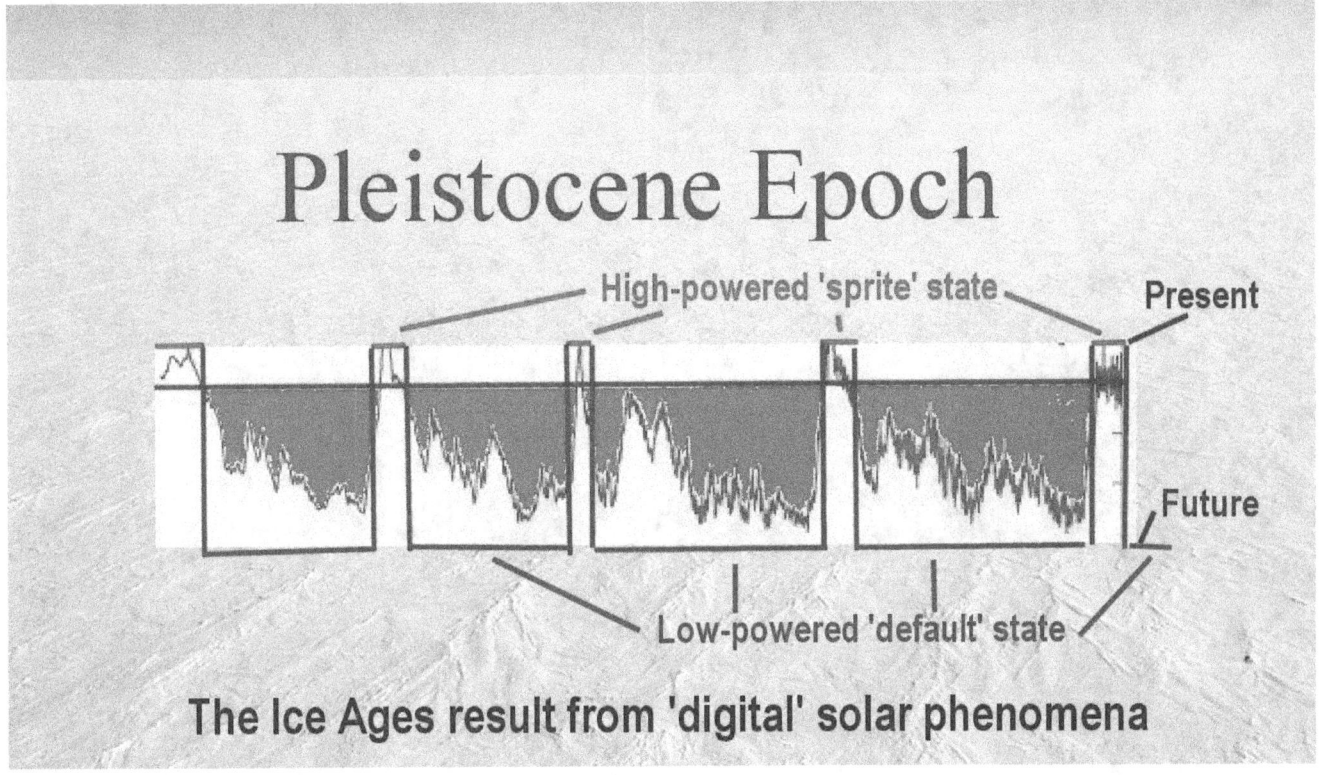

The big ice core records that go back more than 400,000 years tell us the same story with different types of measurements.

But they too, tell us that we are about to fall of the high climate cliff, and that we should develop a landing for it that we can withstand, and that we need to develop this fast, before the event happens, because when the ground gives way under a hiker's boots, things happen almost instantly.

The best that a hiker can hope and pray for, under such circumstances, is for a soft landing. Except we should do better than just hope and pray. We are standing at the cliff already. The ground is rumbling. The rocks are breaking. Whether we will have a soft landing prepared before the ground brakes away, depends on us in the immediate time to create a solution the we know we can live with.

What type of cushion for a soft landing to happen?

NGRIP - The North Greeland Ice Coring Project (1996-2003)

funded by institutions in Germany, Japan, Sweden, Switzerland, France, Belgium, Iceland and the US. Primary sponsor is the Danish Research Council.
(see: http://www.gfy.ku.dk/~www-glac/ngrip/index_eng.htm)

The ice core records cannot answer the most critical question, of what type of cushion we need to develop for a soft landing to happen?

All that the enormous efforts can tell us, which after 7 years produce a single ice core to collect the data from, is that we face an enormous challenge.

This means that the resulting basic recognition is significant only to the extend that it gives us some faint idea of what we might face in the near future when our nicely warm interglacial holiday ends.

The steep climb out of the glacial climate

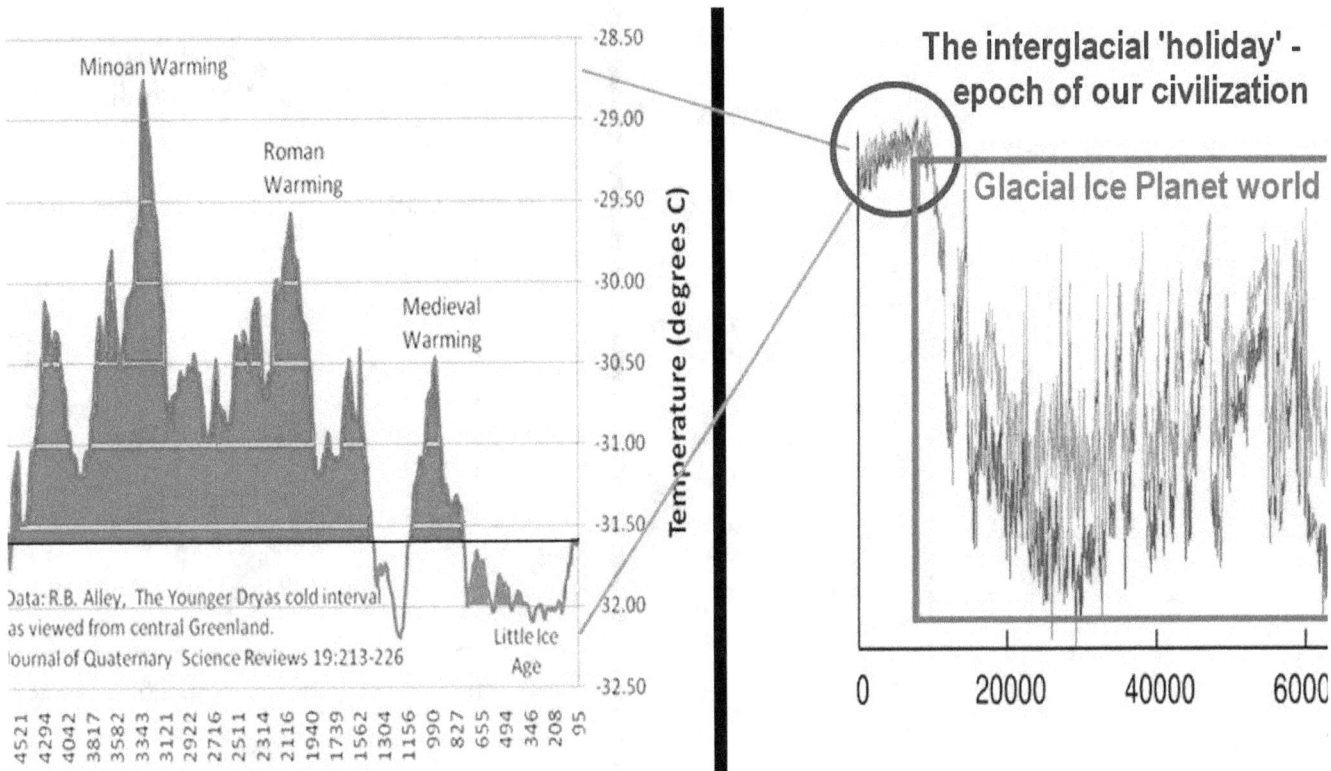

Ice core records can tell us about the steep climb out of the glacial climate to the current interglacial climate that I have encircled in blue, but they cannot tell us what living had been like on Earth before the climb, during glacial times. The best that they can tell us, is that glacial and interglacial worlds are incomparable.

Contained within the narrow band of the interglacial climate

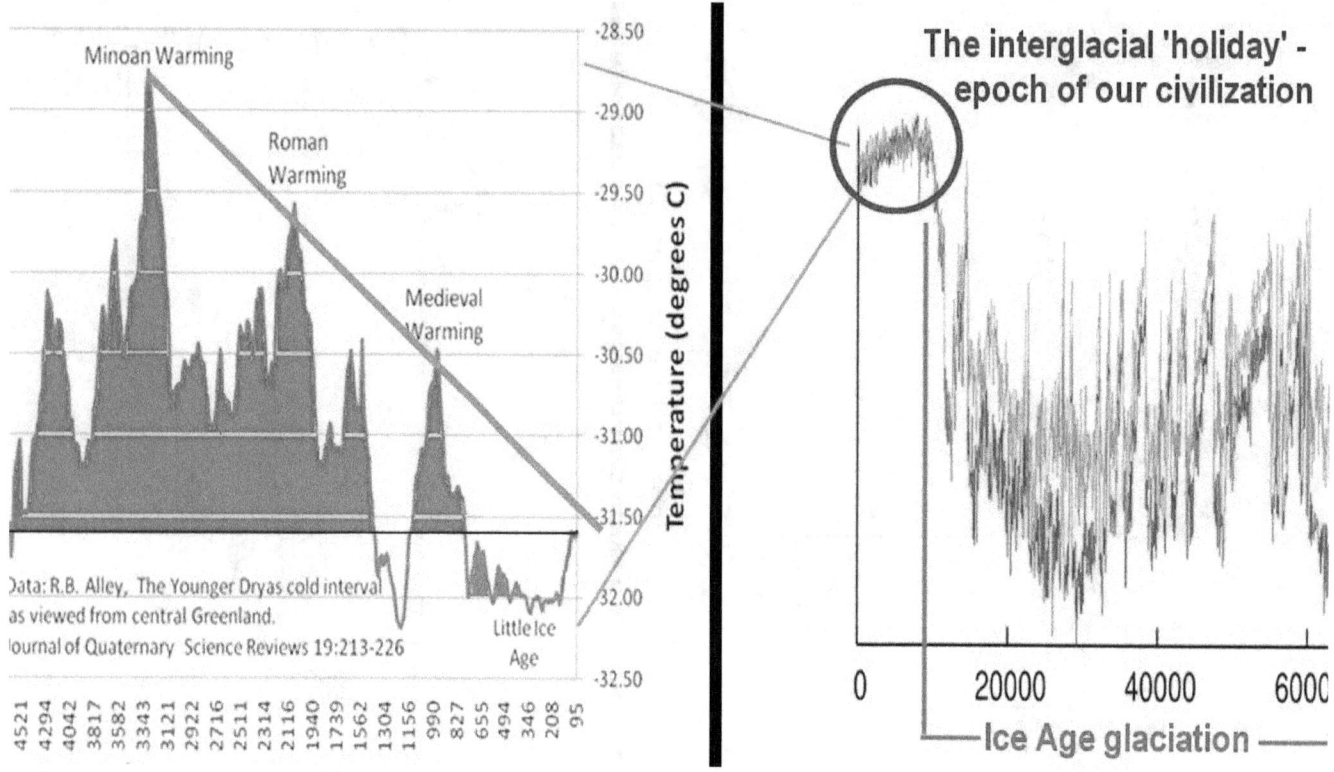

All the climates that we have ever known and experienced, such as the warming periods and the Little Ice Age, are contained within the narrow band of the interglacial climate, within the blue circle. We have no historic records of any kind, from the time before that we can relate to. All that we have, is the measured historic data that speaks of a steep climb to the top, out of deep glaciation, which is about to reverse.

When we fall off the cliff to glacial times

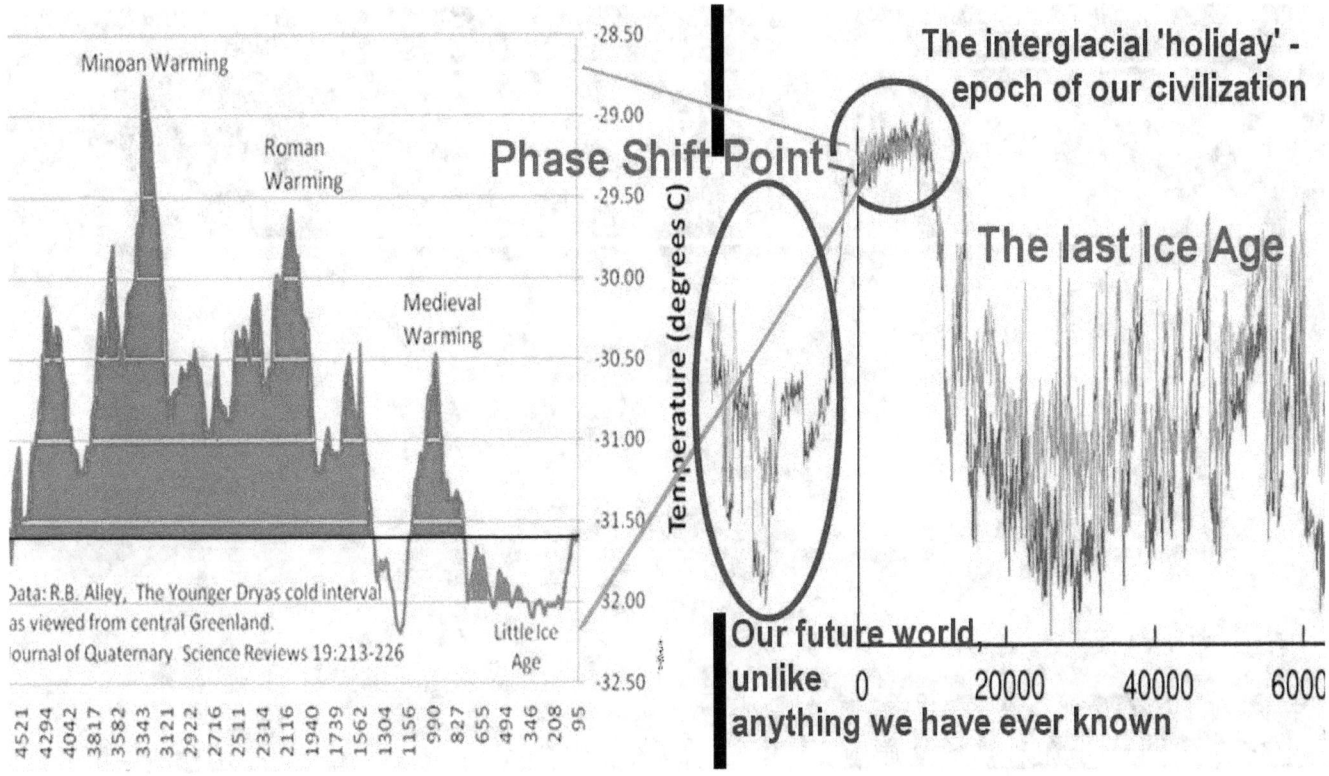

At the very best, the ice core records can give a preview of what we are heading into when we fall off the cliff back to glacial times. This preview is accomplished by projecting forward into our time, the start-up of the last Ice Age, as we have measured it in the North Greenland ice core.

Soft landing for dropping off the high cliff of interglacial

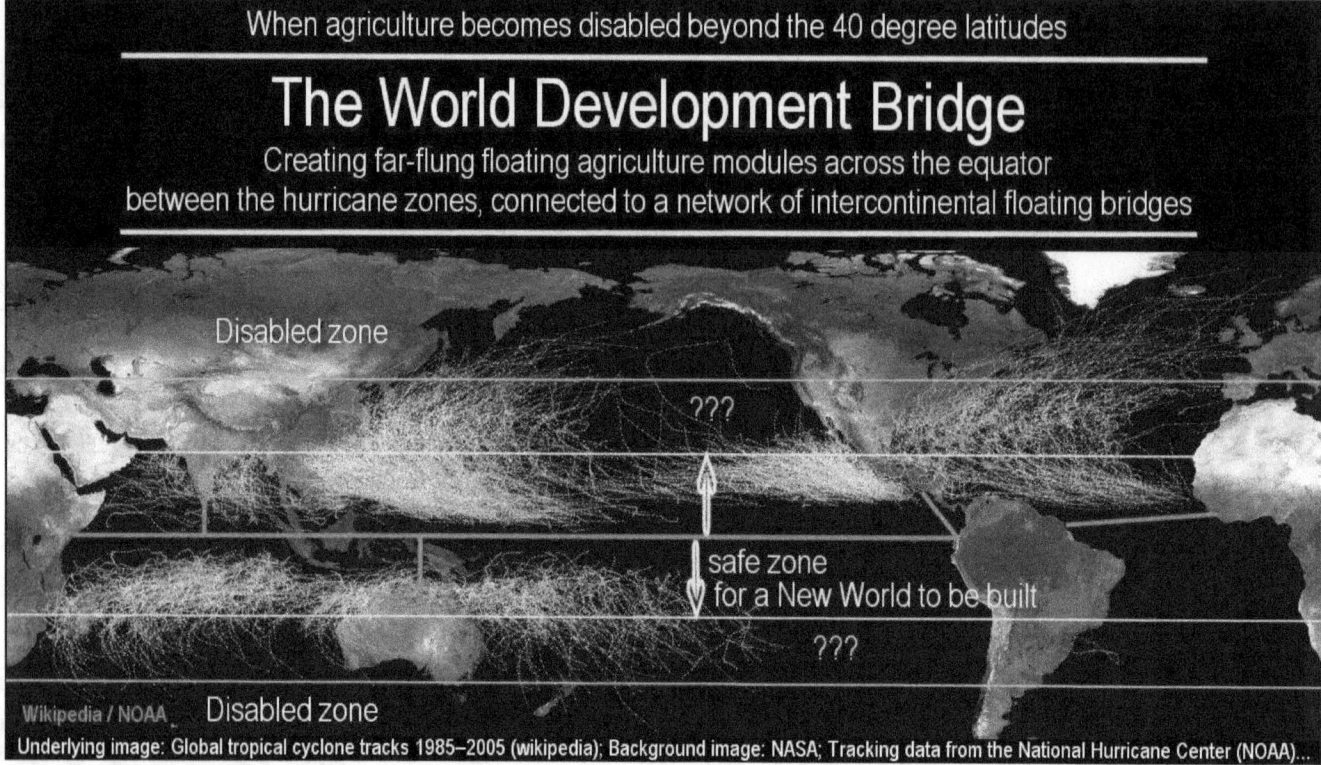

The soft landing for dropping off the high cliff of the current interglacial, requires a substantial cushion. This simply means creating a new world for us that is climate independent, such as in the form of high-tech indoor agriculture with artificial climate and artificial sunlight, which might be strung across the equatorial seas for the lack of suitable land in the tropics.

A more truthful recognition of the nature of the Sun is required

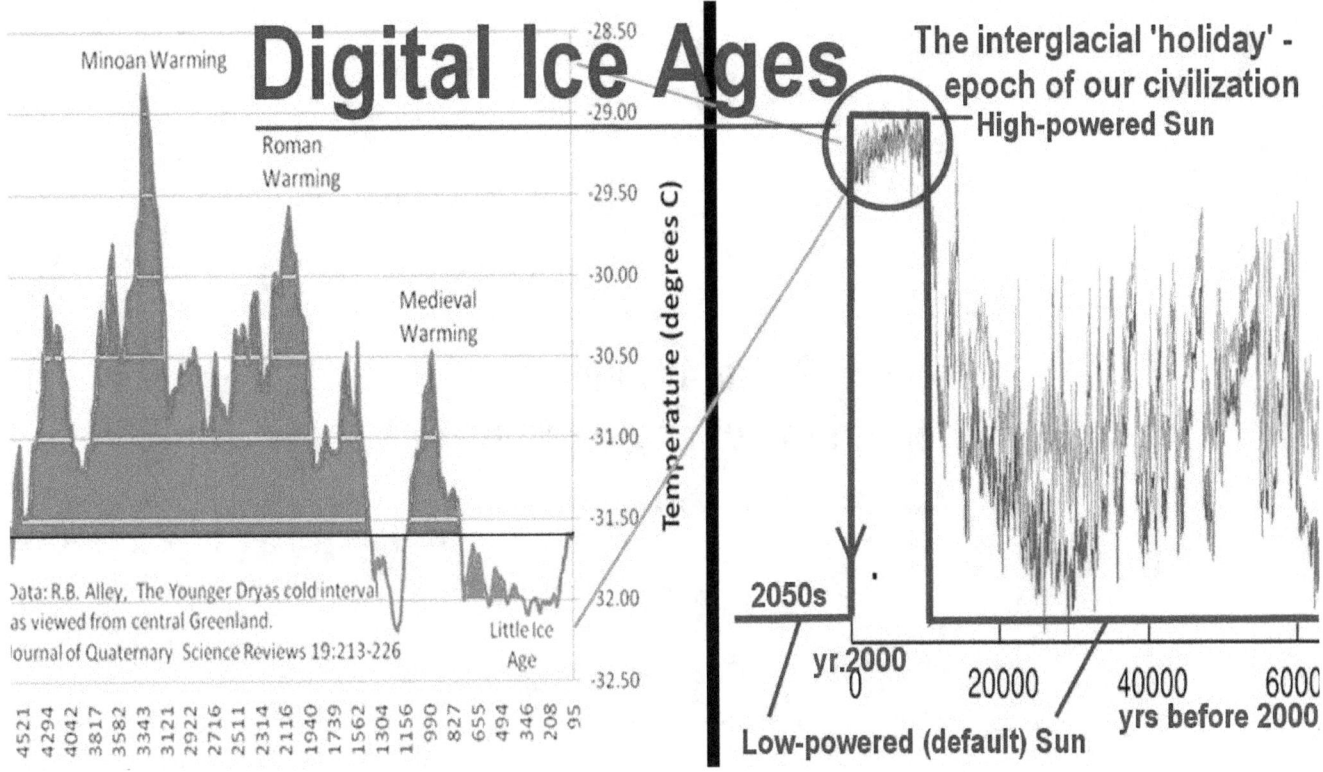

With this kind of 'cushion' set up we have nothing to fear, no matter how deep into glaciation the coming phase shift to the new Ice Age will take us. However, for us to get started with building the cushion, a more truthful recognition of the nature of the Sun is required that is not full of holes and paradoxes, as the mainstream theory is. This, of course, means breaking away from the anti-science movement that H. G. Wells got started.

In order to understand how the dynamics of the Sun can change so rapidly and so immensely, as the ice core records tell us of, further discoveries are needed, and they were made. They were even replicated in laboratory experiments.

# The Plasma Cause

**The Plasma Cause
behind the changing Sun**

The Plasma Cause behind the changing Sun.

Evidence for the plasma cause

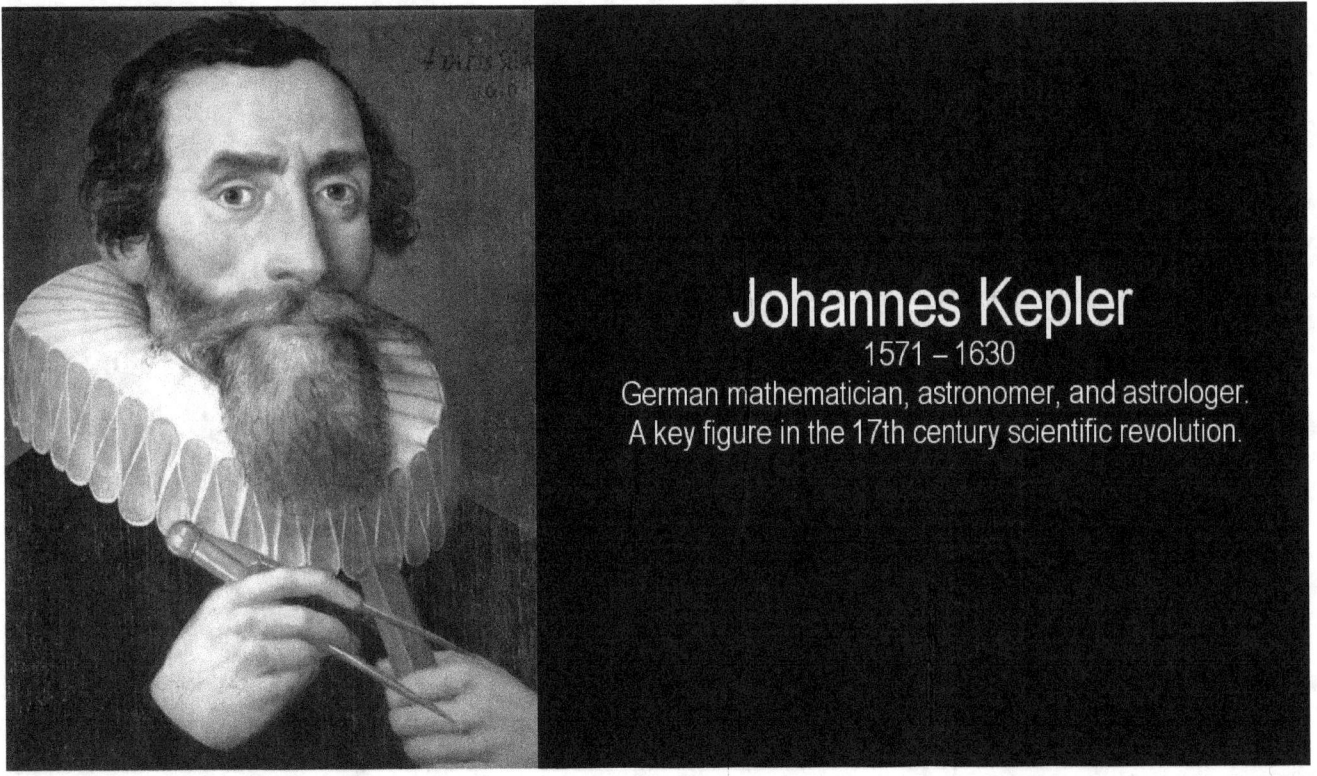

Evidence for the plasma cause that underlies the solar system had already been noted by the astronomer Johannes Kepler, far back in the 1600s.

Kepler was amazed by the geometric progression

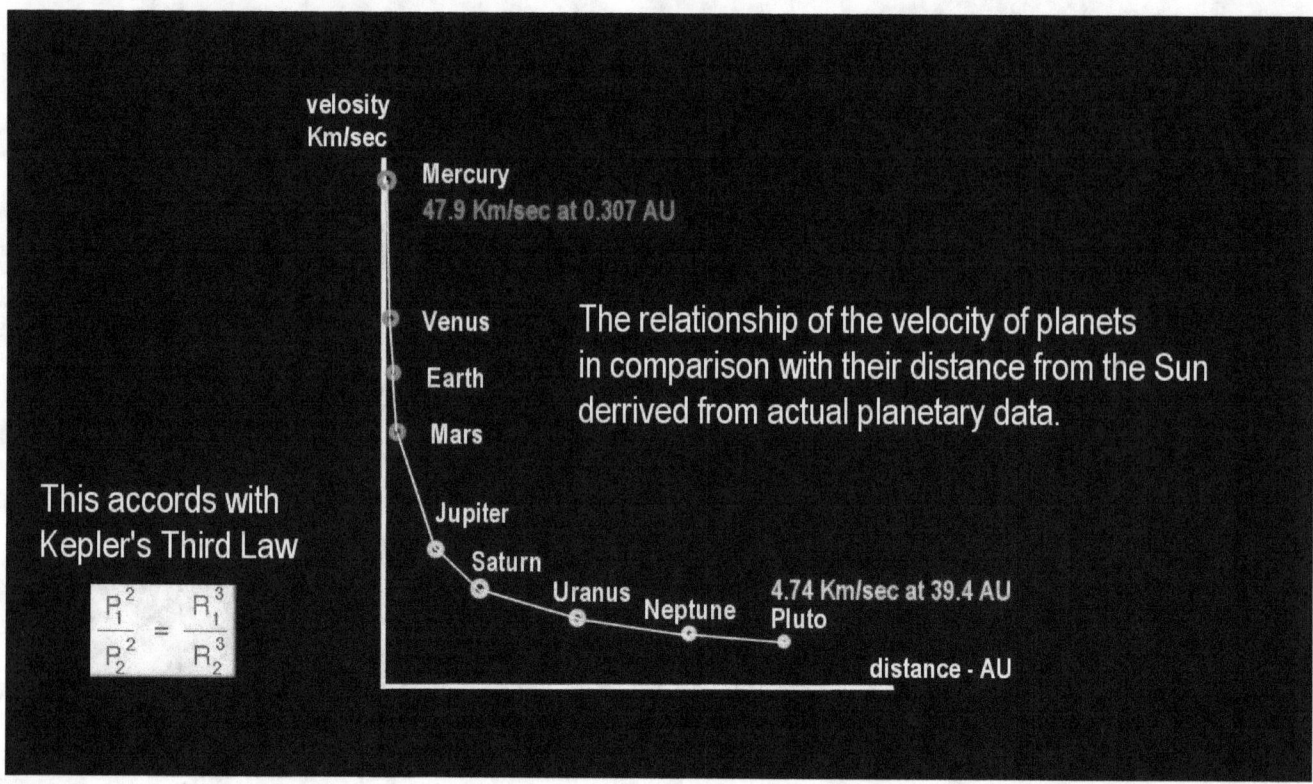

Kepler was amazed by the geometric progression of the spacing of the orbits of the planets, but he couldn't recognize the cause for it, because cosmic plasma physics lay still far in the future.

On the mechanistic platform, no cause can be found

Kepler recognized an intrinsic harmonic relationships in the characteristics of the planetary orbits.

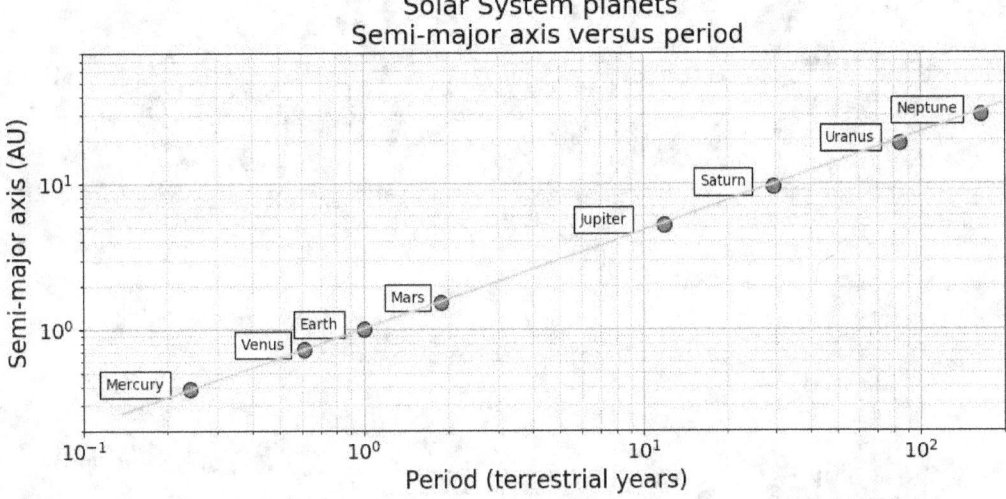

Log-log plot of the semi-major axis (in Astronomical Units) versus the orbital period (in terrestrial years) for the eight planets of the Solar System.

By Mcampestrin - Own work, CC BY 4.0,
https://commons.wikimedia.org/w/index.php?curid=72019799

https://en.wikipedia.org/wiki/Kepler's_laws_of_planetary_motion

On the mechanistic platform, no cause can be found for the orbits' geometric progression, and this with such a precision that their plotting on the logarithmic scale, lines them all up in a straight line.

Kepler was puzzled by the geometric harmony

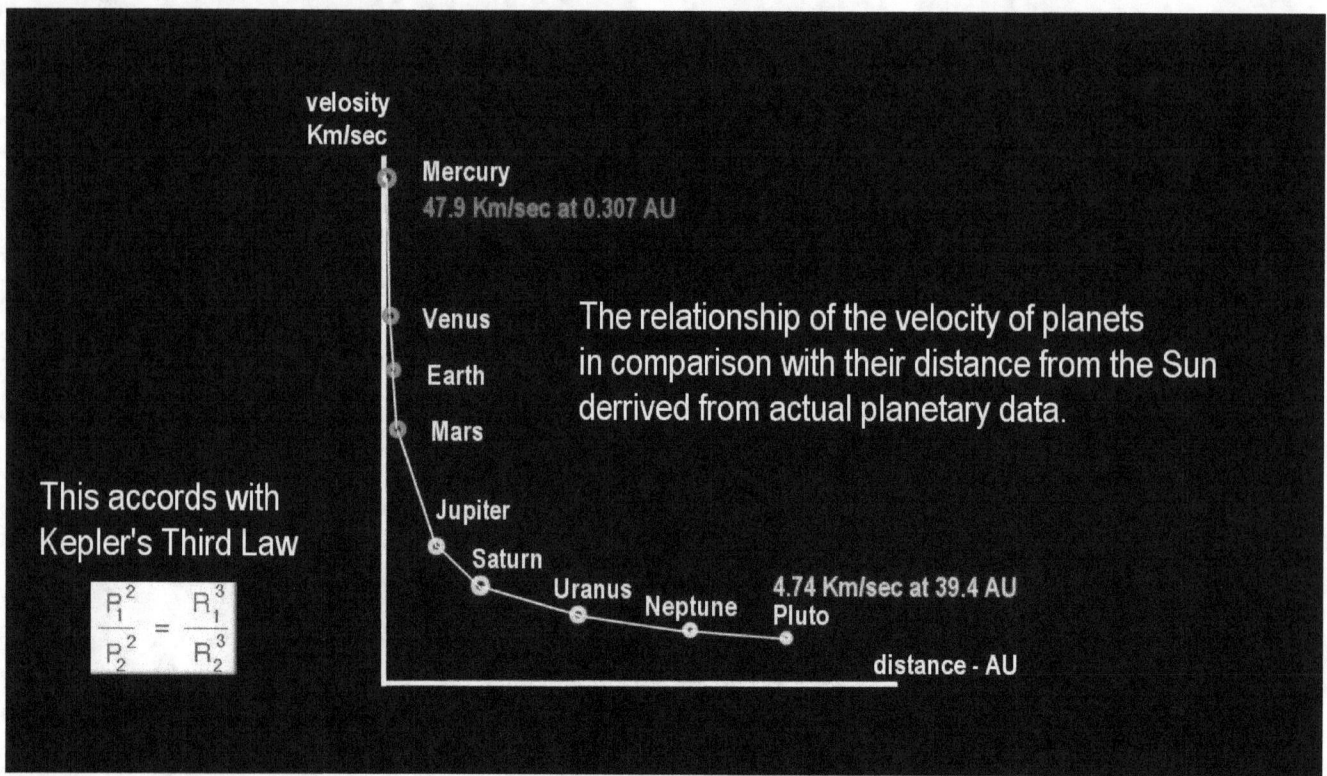

Kepler was puzzled by the geometric harmony of the spacing, where nothing is random, and so he should have been puzzled, because the cause for it can only be found in plasma physics that lay beyond the scope of his vision.

Kepler saw the geometric progression of node rings

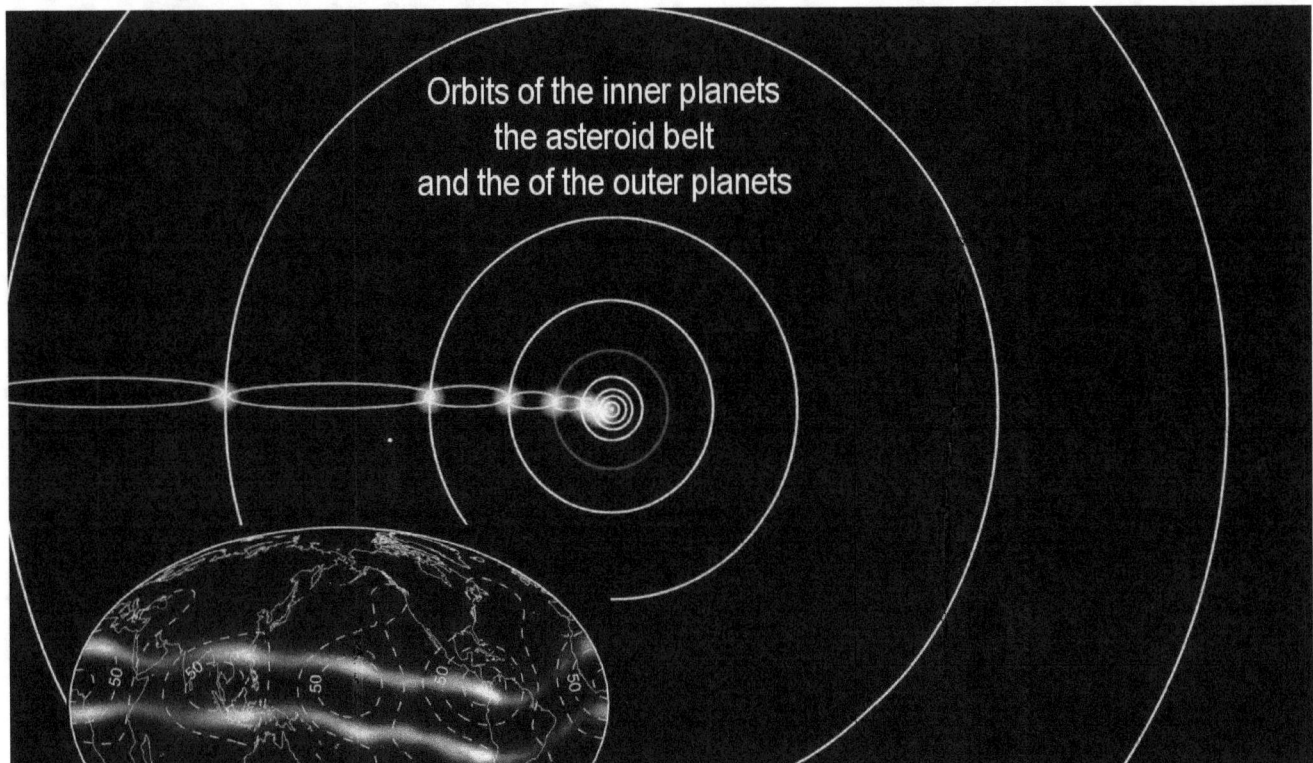

What Kepler saw, was the geometric progression of node rings in the heliospheric current sheet where plasma current is flowing from the heliosphere towards the Sun in a space that becomes geometrically smaller towards the Sun. With the plasma density thereby progressively increasing in the same manner, the resulting node rings in the current sheet occur likewise in geometric progression. In these node rings the orbits of the planets are located, because that's where the planets were formed.

The link with plasma evidence

The link with plasma evidence takes us far out of the mainstream perception of the solar system. Flowing plasma in space is deemed not to exist, but numerous types of evidence of this sort tell us that plasma in space does exist.

Plasma in Space

a brief overview.

Plasma in Space a brief overview.

The Sun itself is a body of evidence

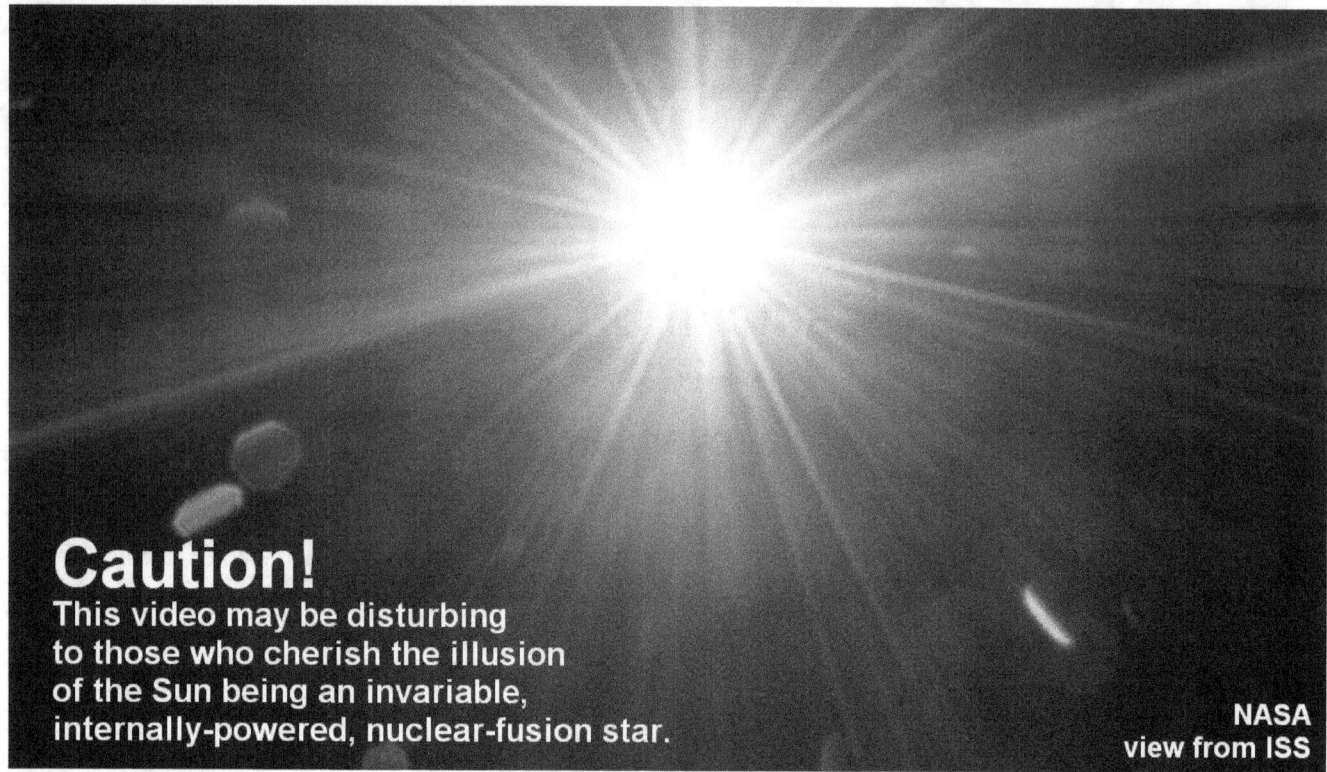

The Sun itself is a body of evidence, with its numerous features, that reflect plasma principles. While plasma itself is invisible, with its protons being 100,000 times smaller than an atom, and its electrons being 1,000 times smaller than that, we can see the plasma principles in operation in the effects they create, especially those that affect us, like the solar wind and the solar cosmic-ray flux, and also the Sun's radiation in light and heat. From these effects we can discover the nature of the principles by which the solar system operates. This is critical for our orientation towards the near future.

So, what are its principles that we can detect by their operation?

Plasma is made up of electric particles. In their motion, magnetic principles apply

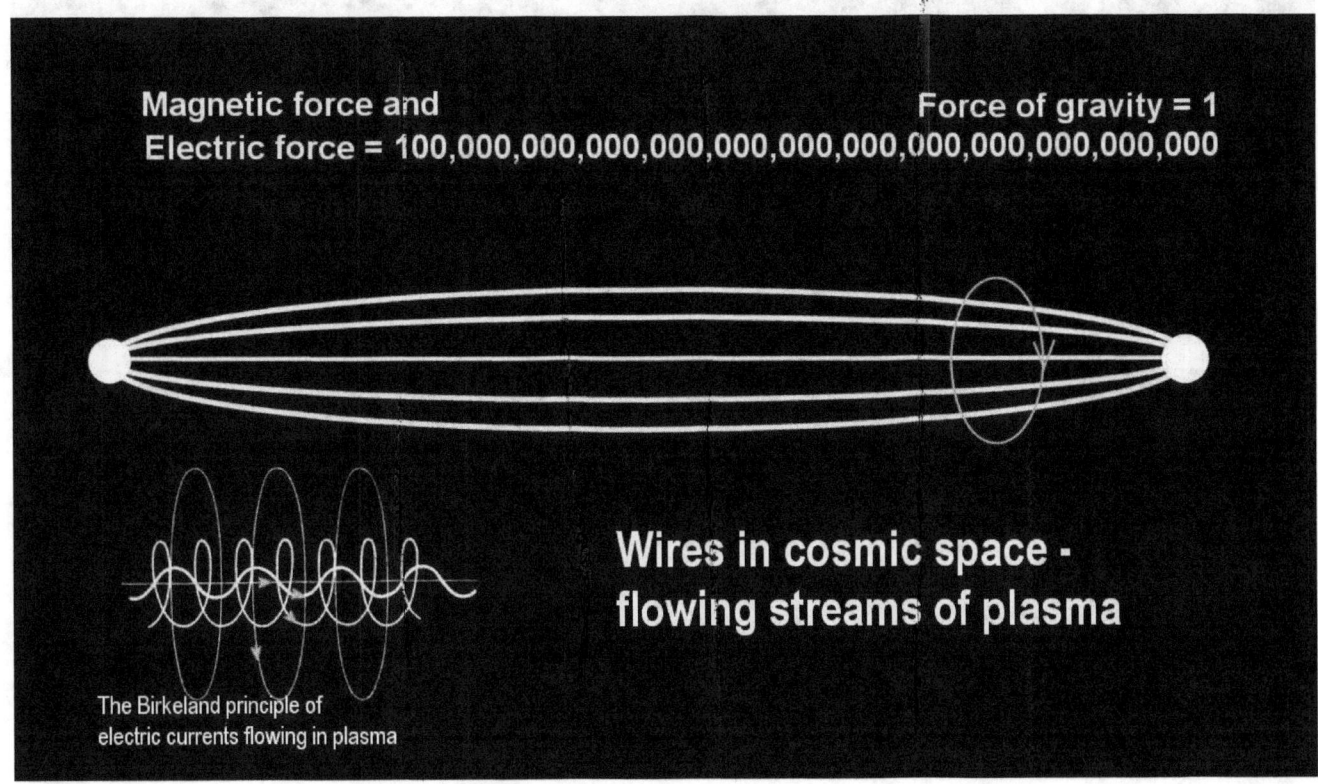

Plasma is made up of electric particles. In their motion, magnetic principles apply in addition.

When Electric currents flow in parallel wires in the same direction, the flowing electricity creates magnetic fields by electrons in motion. The interacting magnetic fields draw the wires to each other, by what is termed the Lorenz Force.

In space plasma flows in streams

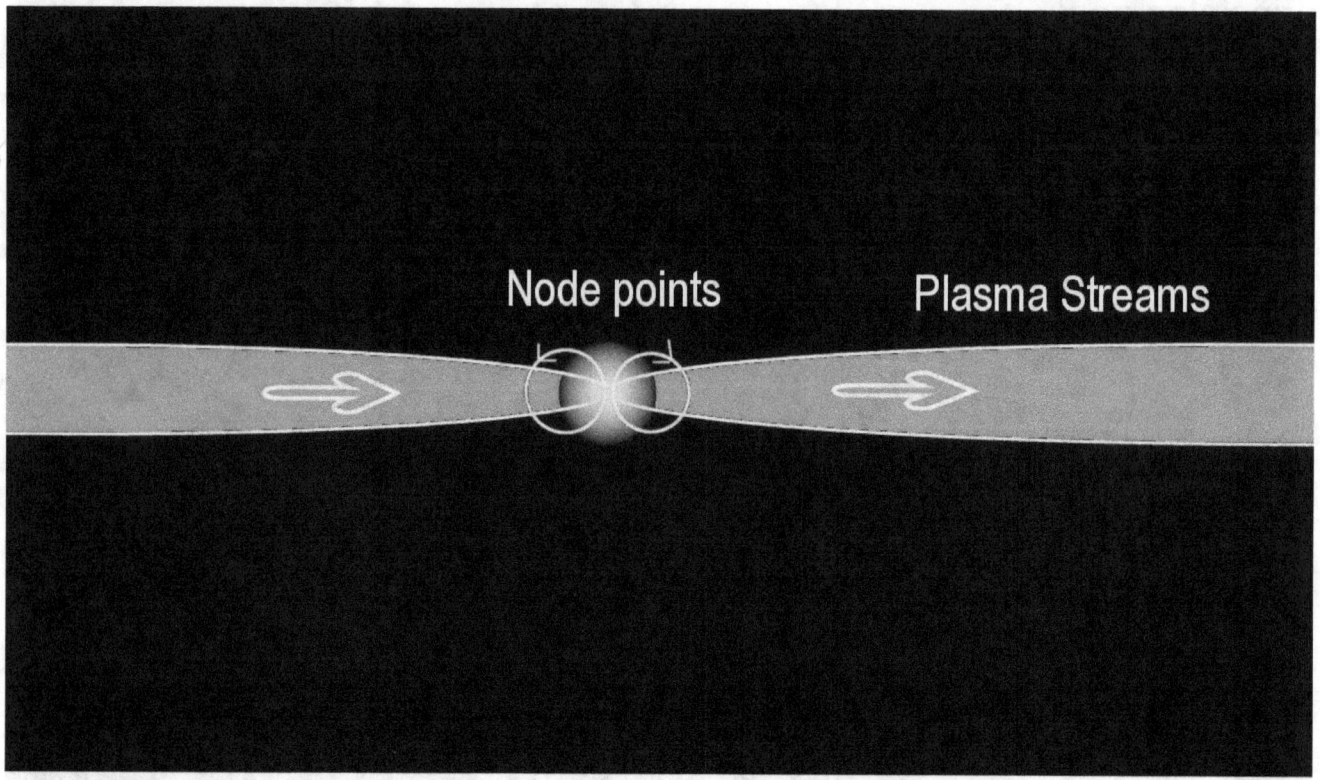

In space, where plasma flows in streams without wires, the developing magnetic fields pinch the entire stream together into a smaller cross-section, which thereby increase the current density and and magnetic field strength, which increases the pinching still further. At the extreme, the magnetic fields tangle up and form a node point, where typically a star is located, like our Sun.

When the magnetic fields tangle up

based on
David LaPoint: The Primer Fields

When the magnetic fields tangle up, they force the plasma to flow backwards and under a magnetic confinement dome, where the plasma becomes concentrated, and from there, flows in concentrated form onto a Sun.

With the Sun being a sphere of plasma itself, with an electron-rich high density surface, the inflowing concentrated plasma interacts with it. In the resulting extremely dense process, a portion of the plasma becomes fused into highly energized atomic elements that radiate light and heat. The synthesized atomic elements, by them being electrically neutral, gradually flow away from the Sun with the solar wind.

The remaining plasma, that is less dense now, also flows away from the Sun. It flows away in the reverse of the process. In flowing away, it expands again into an interstellar plasma stream that flows on to the next star and collects more plasma from space along the way.

David LaPoint named the magnetic fields, The Primer Fields

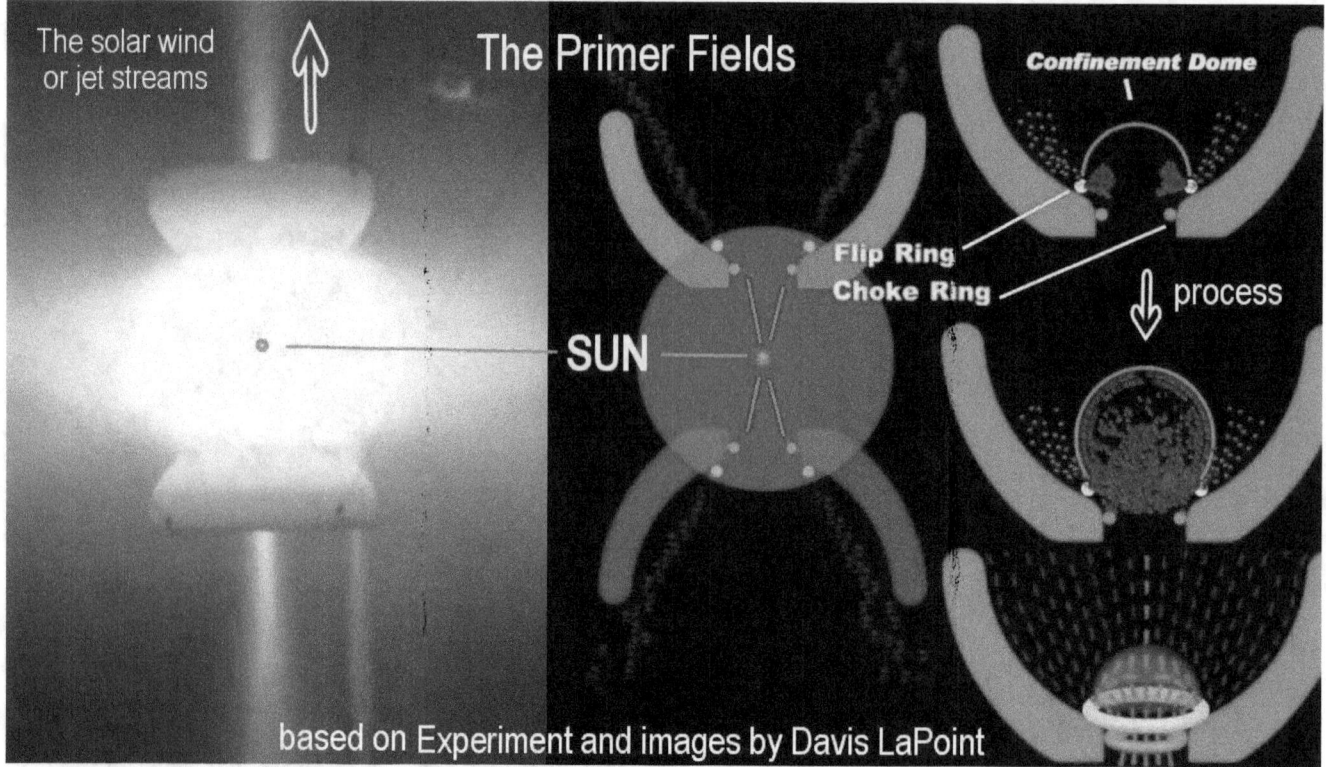

The researcher, David LaPoint, who explored the principle in laboratory simulation, named the magnetic fields, The Primer Fields. He also explored the principle of the solar wind that is vented explosively from the top of a confinement dome.

He replicated in laboratory experiments

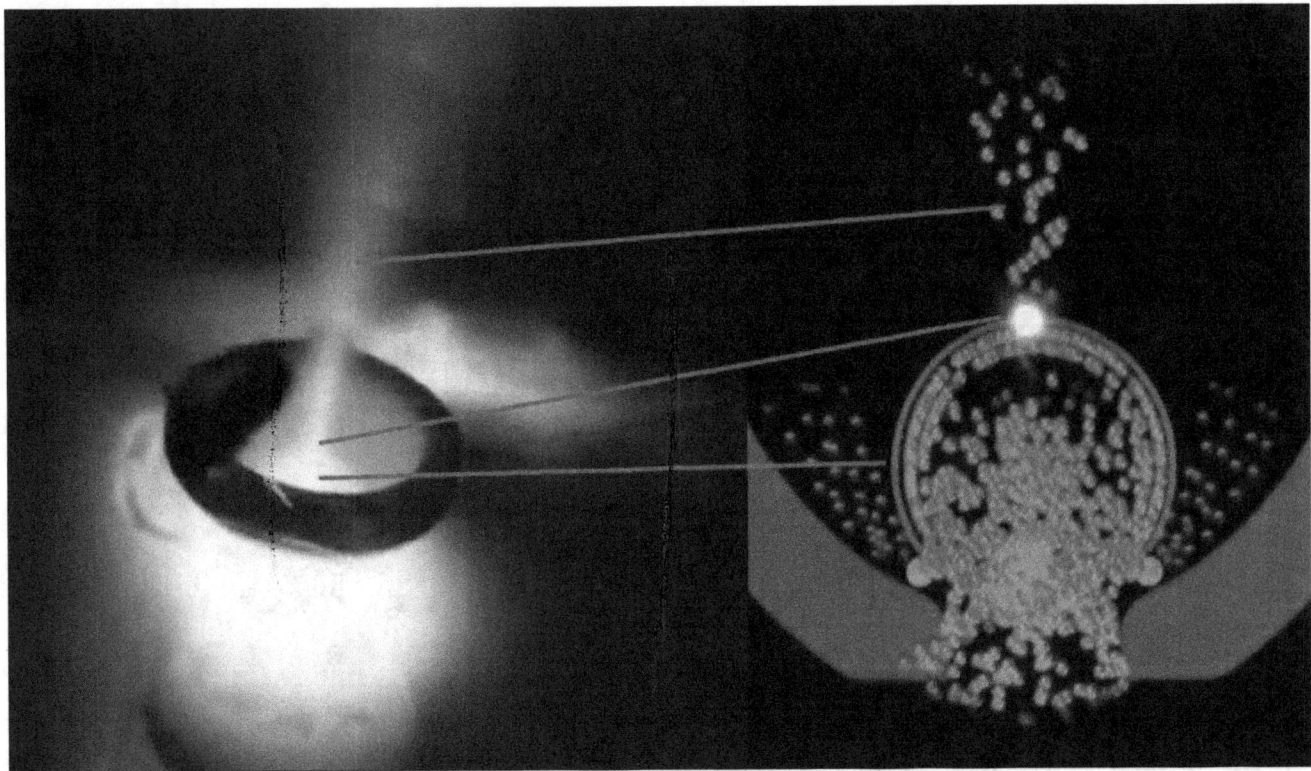

He replicated in laboratory experiments, that when the pressure under the dome exceeds the confining magnetic field strength, the excess pressure is vented, which on the surface of the Sun becomes the solar wind. While the solar wind doesn't affect the climate on Earth, the measured rate of diminishment establishes a benchmark for predicting when the solar wind will stop, because any further weakening of the input plasma stream, will diminish the concentration of the plasma that is focused unto the Sun, by which the Sun will dim and become cooler.

That's when a new phase of the weakening of the solar activity begins.

Continued weakening will collapse the the Primer Fields

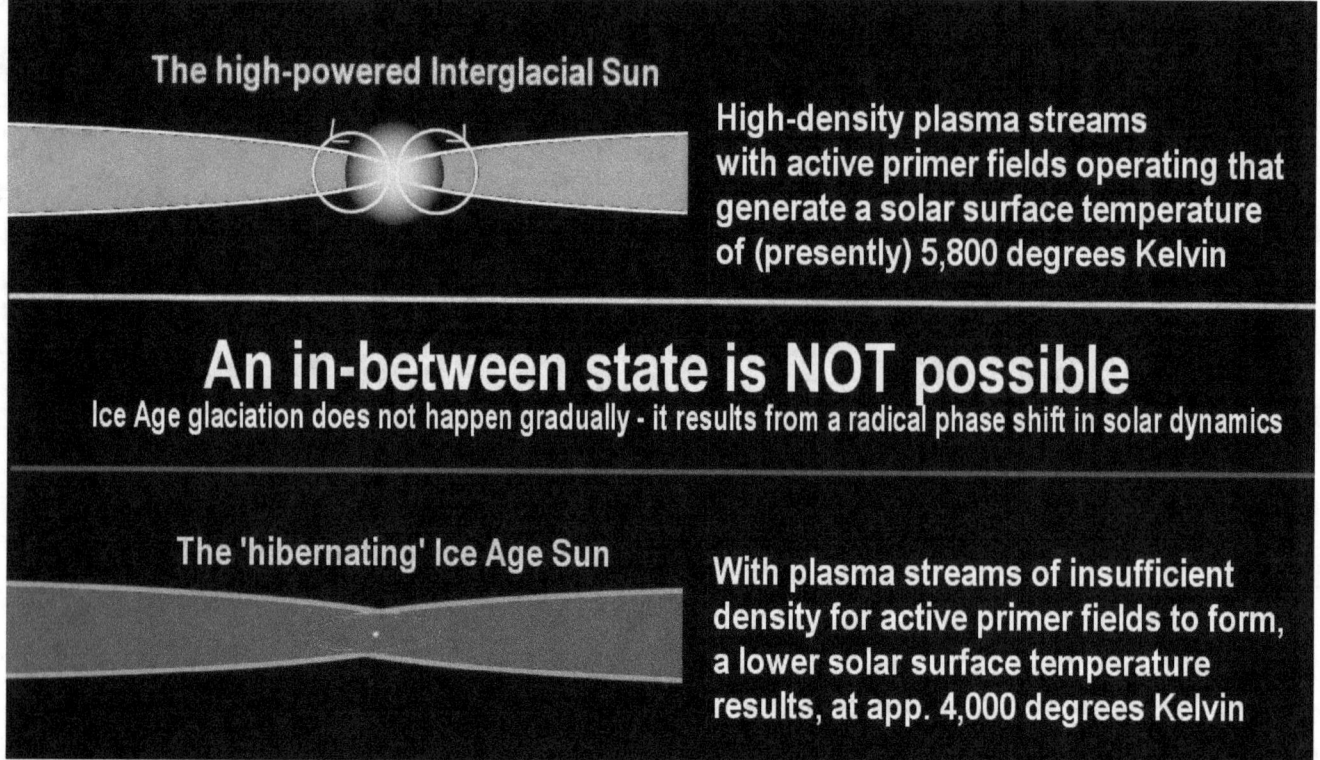

The continued weakening of the input plasma stream, will eventually collapse the the Primer Fields. When this happens, the interstellar plasma stream will still keep on flowing, but it will flow loosely by the Sun, instead of being focused onto it.

At this point, the Sun drops into a low-power mode and hibernates till the interstellar plasma stream recovers again in plasma density. That's how the Ice Age phase shift happens. We will get to this point, potentially, in the 2050s.

The bottom line is, that Ice Age conditions are created by a totally different process than the climate weakening that is presently in progress in the early part of the boundary time zone to the Ice Age phase shift.

I have explored the dynamics extensively

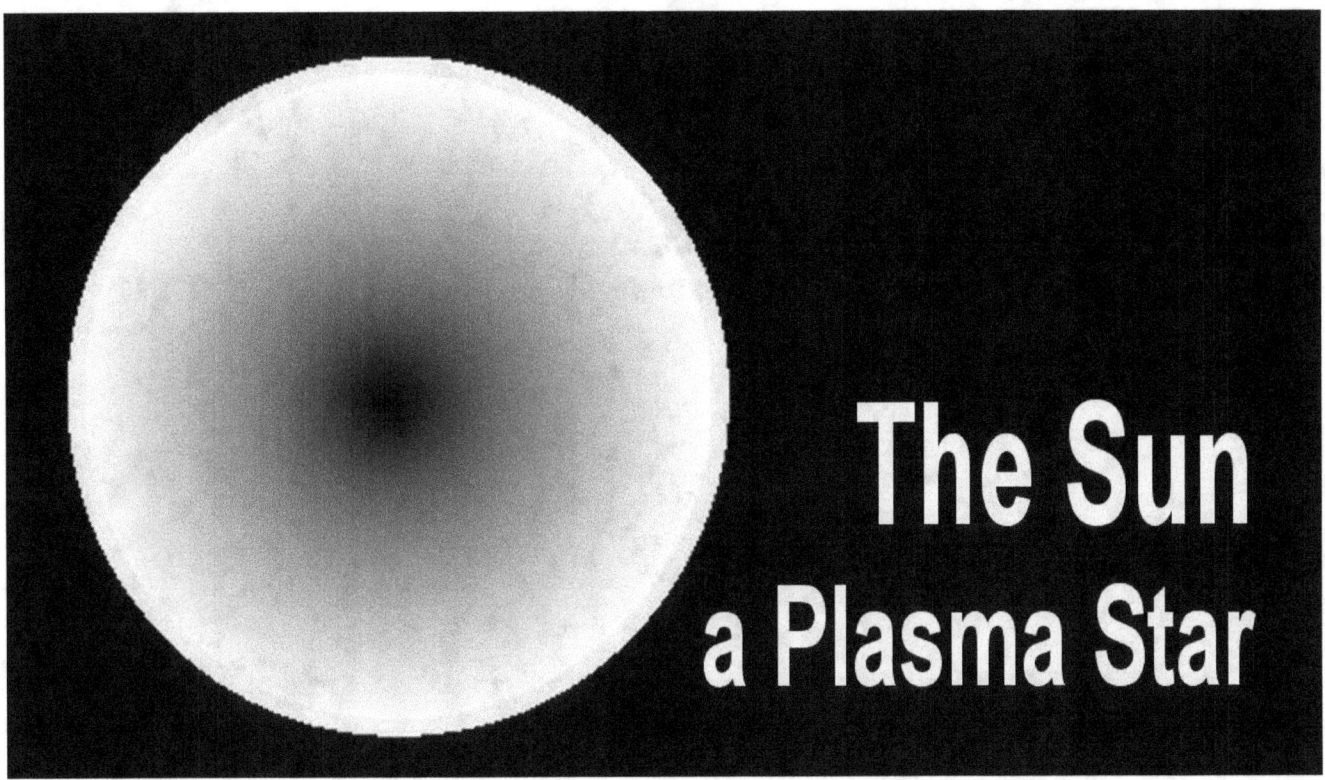

I have explored the dynamics of these discoveries extensively over the years and presented them in my numerous exploration videos, because an understanding of the principles involved, by which the Sun is powered and is affected, enables us to understand the digital nature of the Ice Age cycles, and to forecast the timeframe in which the next phase shift to glaciation will likely occur.

Another mayor contributor to the Primer Fields theory

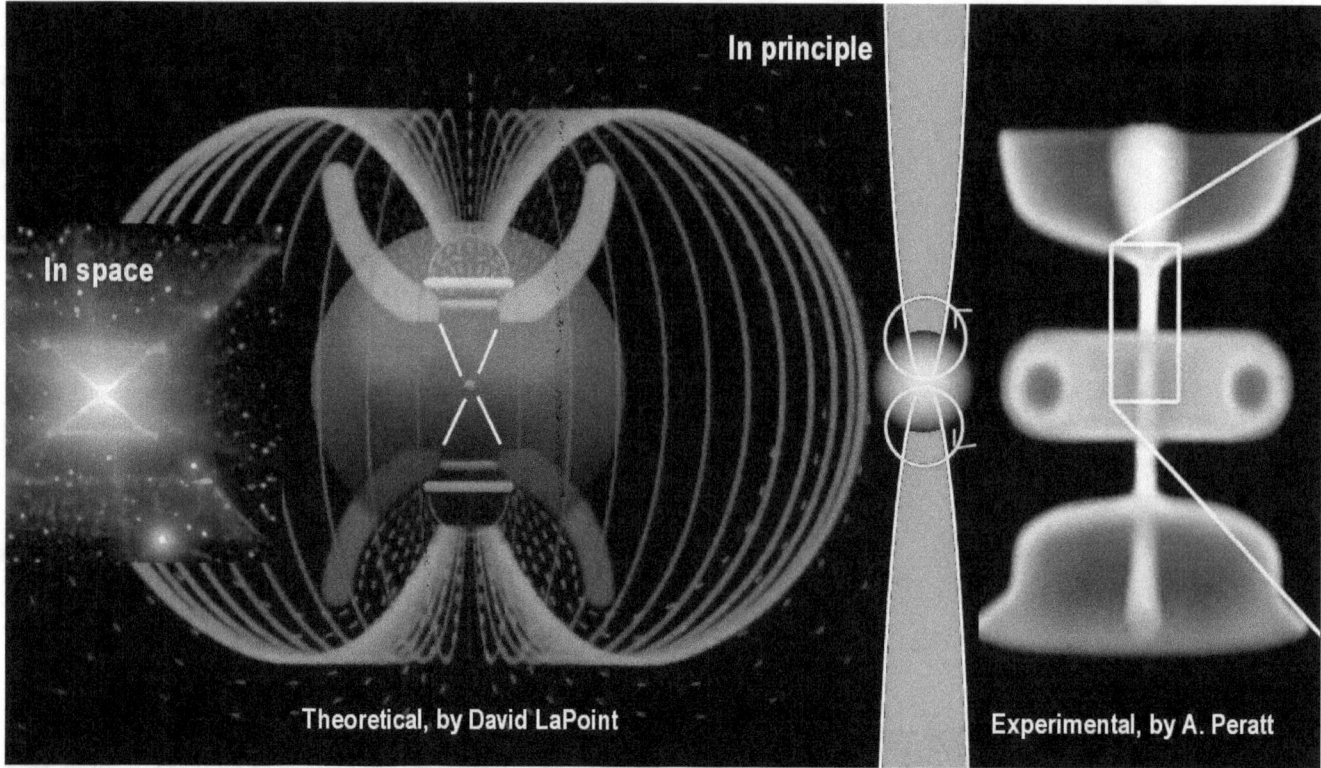

Another mayor contributor to the Primer Fields theory, was the director of experiments at the Los Alamos National Laboratory, Anthony Peratt. He achieved the experimental replication of the dynamic forming of the Primer Fields, which David LaPoint had explored in static simulations.

An understanding of the dynamics of the principle involved, is critical for all humanity, so that it will rouse itself out of its easy chair and start building itself the technological cushion that enables a soft-landing into the next 90,000-years glaciation period that potentially begins in the 2050s with the collapse of the Primer Fields

Another major contributor was the NASA and ESO Ulysses satellite

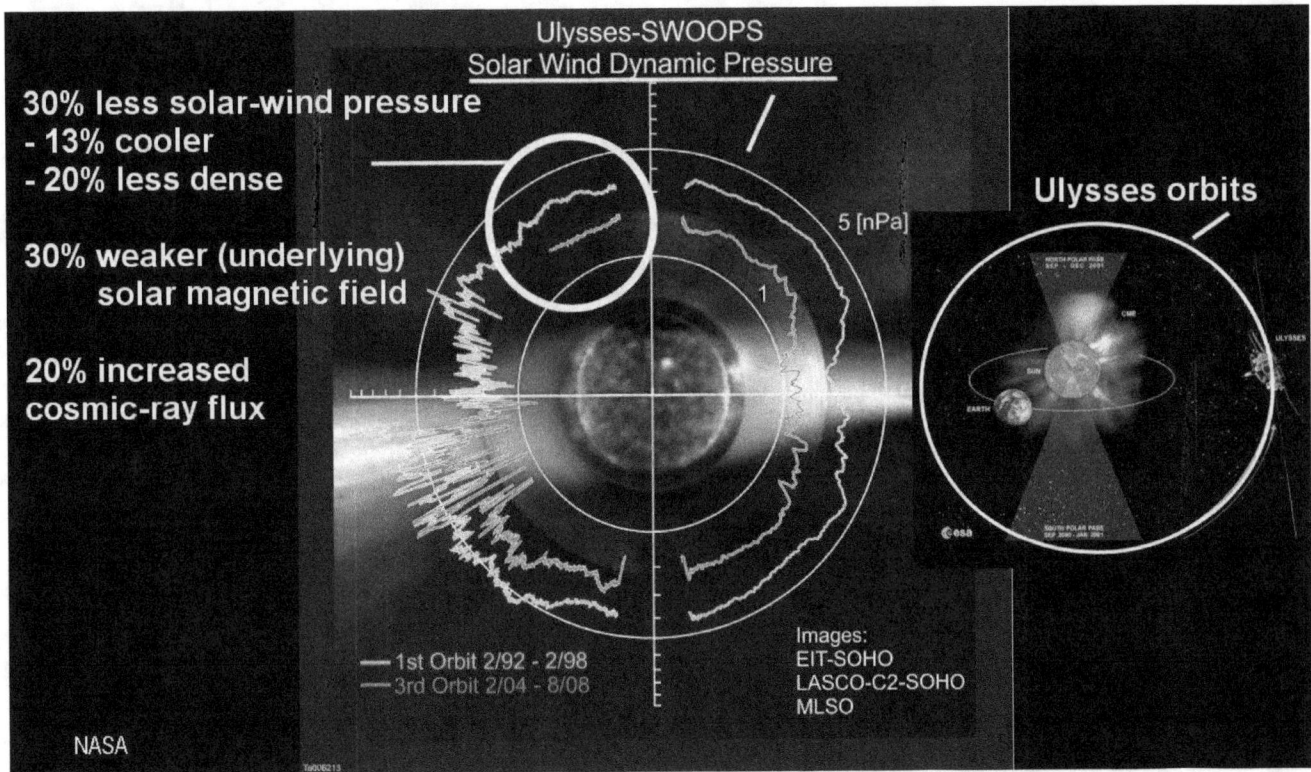

Still another major contributor for understanding the timing of the dynamics that are in progress, was the NASA and ESO Ulysses satellite. The satellite's measurements brought home 3 breakthrough discoveries. We saw in the measurements the solar-wind pressure collapsing by 30% over the span of a decade of its operation, and we saw the solar cosmic-ray flux increasing by 20% over the same time span. These measurements have established a benchmark, both for the starting point and for the rate of the collapse that still continues, as subsequent measurements indicate.

The voids in the solar wind over the poles of the Sun

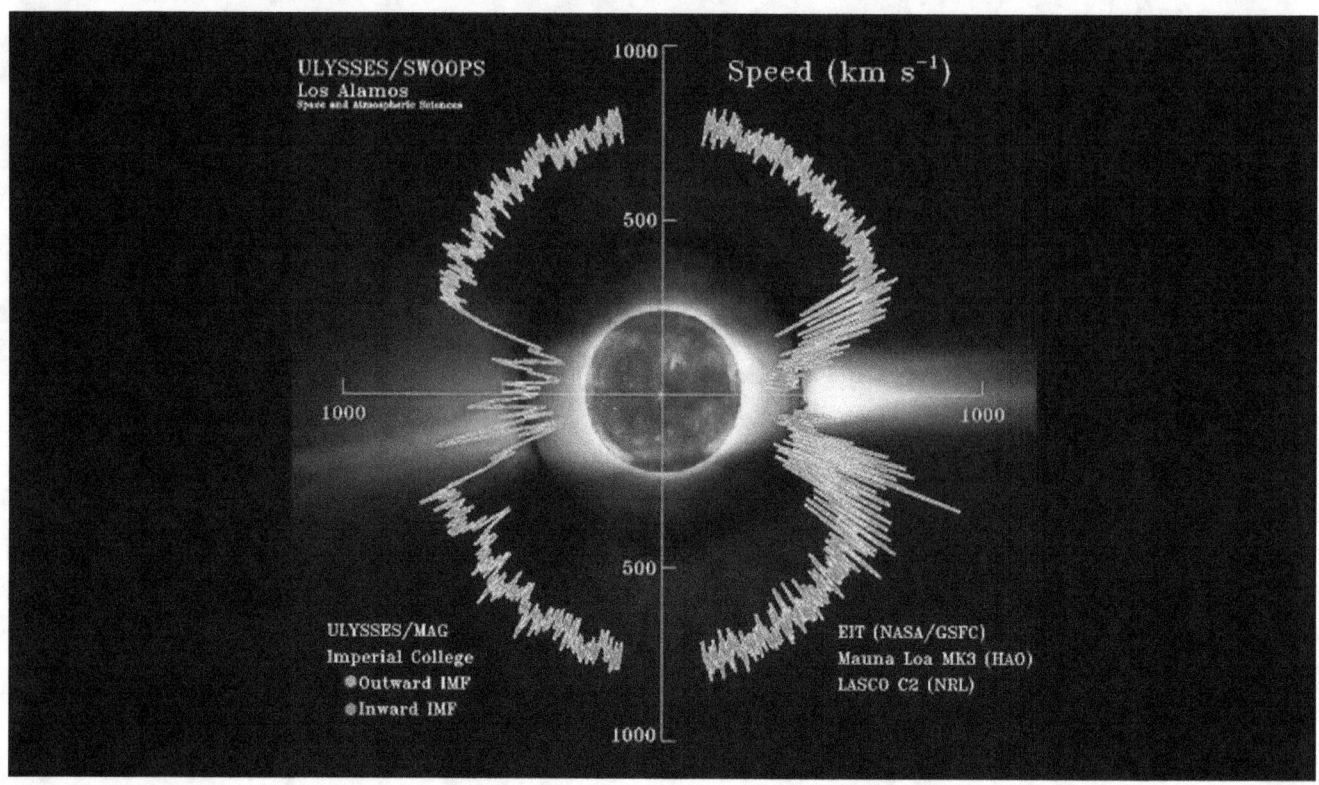

The third, major discovery that Ulysses measurements provided, is the discovery of the voids in the solar wind over the poles of the Sun.

The Ulysses mission certified the Plasma Sun

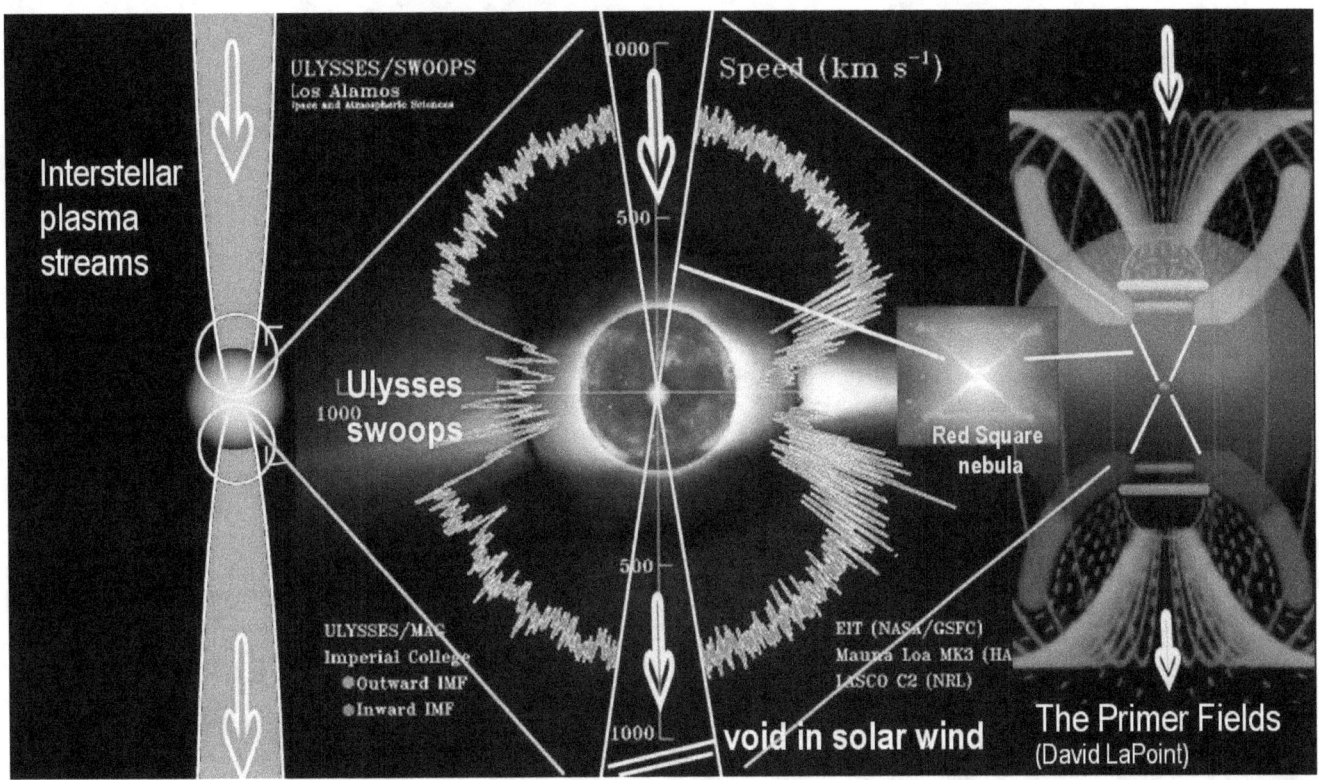

With the discovery of the voids, the Ulysses mission certified the Plasma Sun concept as an established fact.

The measured void provides evidence that the Sun is indeed a Plasma Star that is powered by a highly concentrated stream of plasma being focused onto it by its Primer Fields, which experiments have indicated would flow unto the Sun over its polar regions.

Ulysses about the solar wind

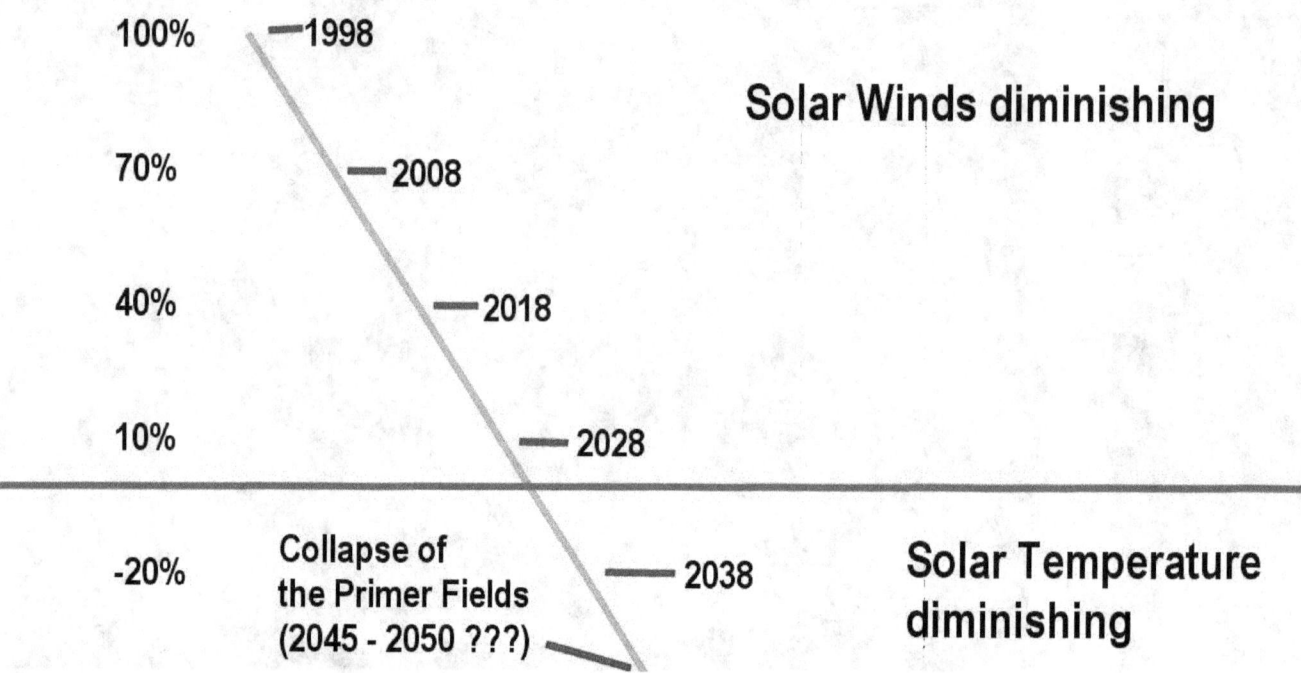

So let's look at what Ulysses tells us about the dynamics of the diminishing solar wind.

The measured rate of diminishment of the solar wind pressure has opened for us a way to more fully understand of the dynamics of the Ice Ages.

Yes! If one projects the weakening of the solar-wind pressure forward, at the measured rate of 30% per decade, the solar wind will stop in the 2030s. This is not only possible, but is actually expected due to the nature of the solar wind.

The solar wind no longer occurs

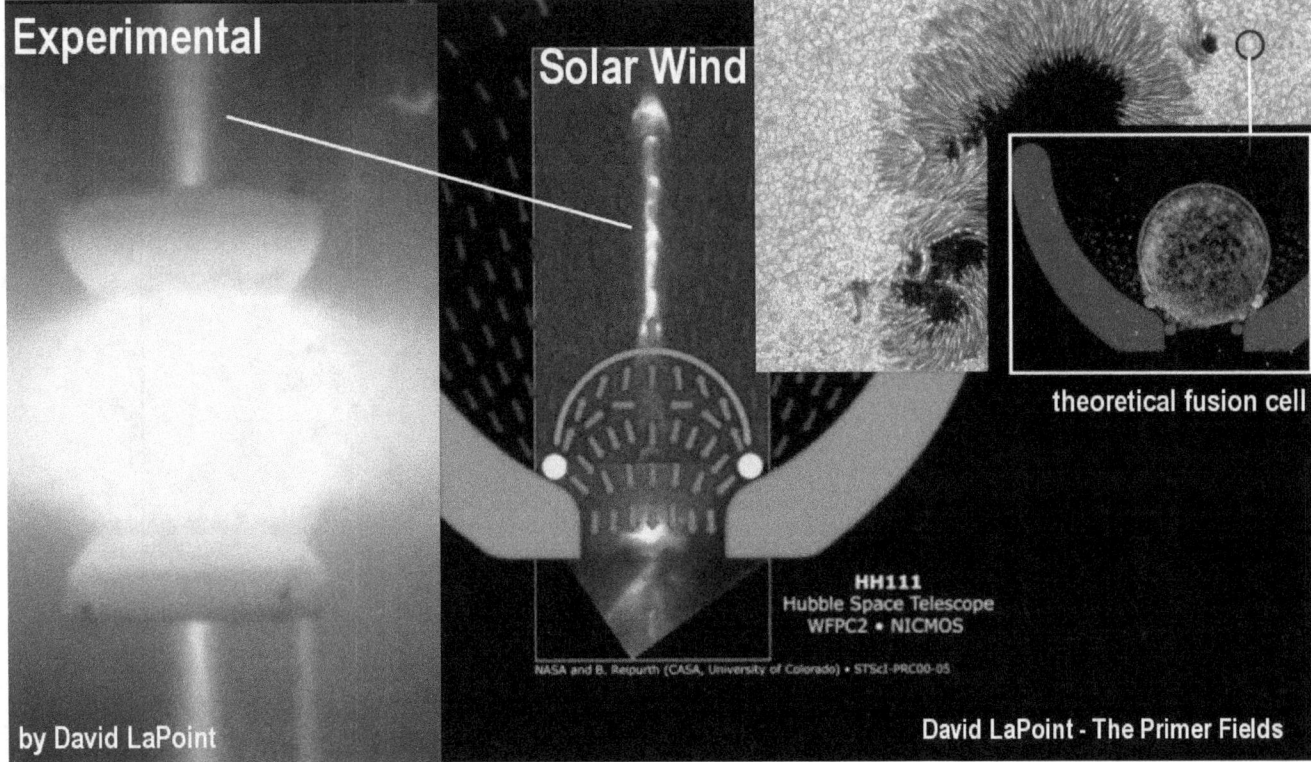

By the magnetic dynamics of the Primer Fields, plasma is curled backwards at the node point, under a magnetic confinement dome where it becomes concentrated. As I said before, when the resulting pressure exceeds the magnetic strength of the confinement field, some of the plasma escapes at the weakest point, at the top of the dome.

The escaping plasma, from the top of the fusion cells, on the surface of the Sun, becomes the solar wind. As the density of the inflowing plasma stream diminishes, the solar-wind density diminishes with it. It will diminish until the pressure under the dome no longer exceeds the confining field strength. At this point, the solar wind no longer occurs.

While the solar wind flows

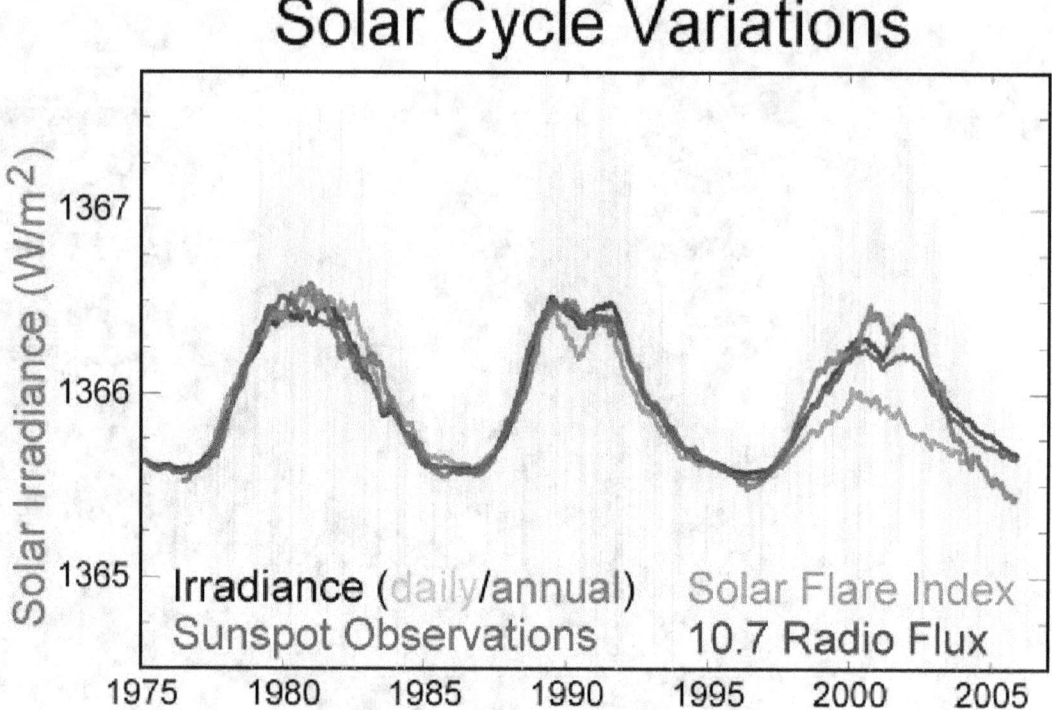

While the solar wind flows, the solar fusion process is only minutely affected by the continuously weakening interstellar plasma stream. The radiation doesn't vary much in this case, because the pressure under the dome doesn't vary much. The magnetically controlled venting of the solar wind keeps the pressure precisely regulated.

The regulated pressure, is the pressure that is focused onto the Sun where atomic elements are synthesized that emit light and heat. For as long as the pressure continues to be regulated, by excess pressure being drained away as solar wind, the solar radiation remains steady as a rock, within a fraction of a percent.

However, when the input stream weakens to the point at which solar wind is no longer ejected, and the input stream continues to diminish further, in density, then the reduced plasma pressure under the dome, is the pressure that is focused onto the Sun, whereby the fusion process diminishes accordingly, and the surface of the Sun becomes colder.

The cut-off of the solar wind is critical as an indicator

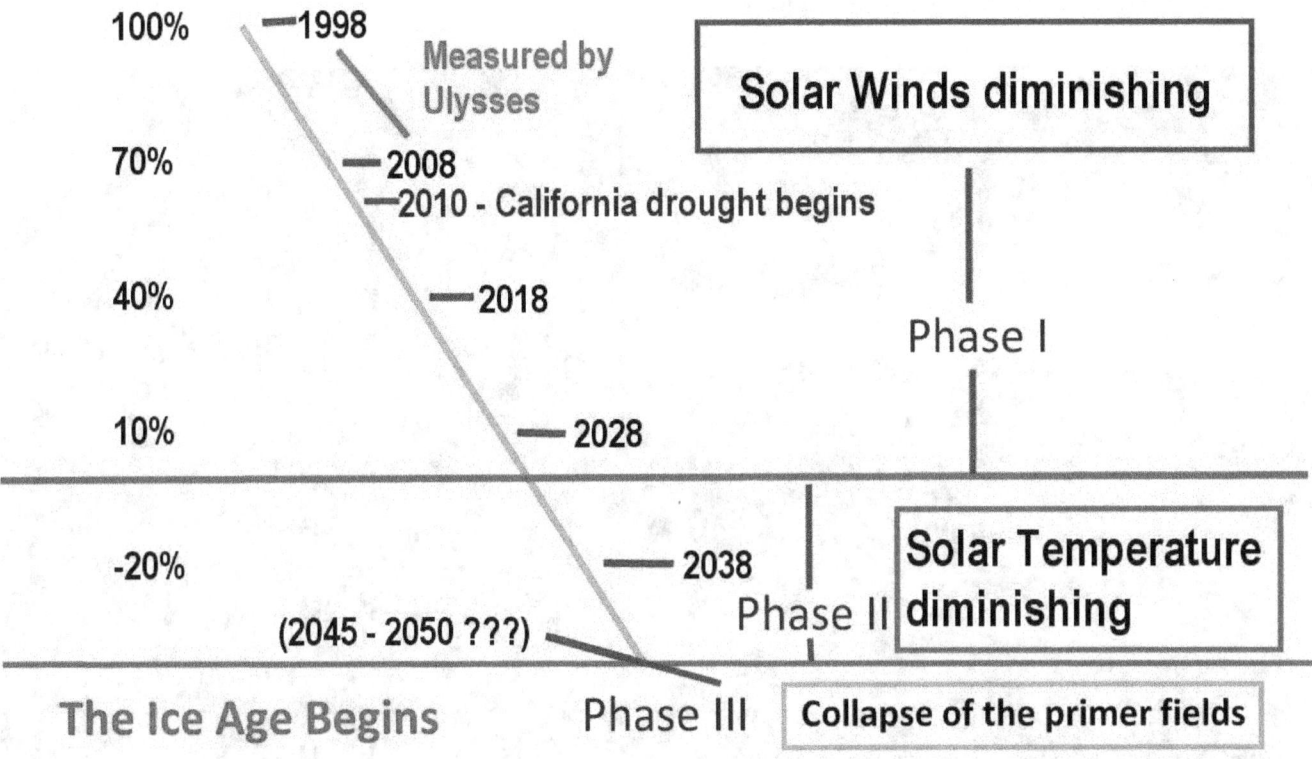

The point of the cut-off of the solar wind is critical for us as an indicator, because that's when Phase 1 of the Ice Age start-up sequence, ends and Phase 2 begins.

We will get to this point in the 2030s, at the present rate of diminishment of the solar wind. Up to this point, only the increasing solar cosmic-ray flux affects our climate. Past the point of the solar-wind cut-off, the Sun begins to dim, which then also affects our climate, making it still colder.

Phase II is the phase of the diminishing solar radiation. For how long this phase will last is hard to predict. The collapse process appears to be self-amplifying.

A dimmer Sun consumes less plasma. The reduced plasma consumption reduces the rate of flow through the Primer Fields. Because the Primer Fields are dynamically created magnetic structures that are formed by the effects of flowing plasma, a stage will be reached when the diminishing rate of flow is no longer sufficient for the Primer Fields to be maintained. When this happens, the Primer Fields will collapse. They will vanish as if they never existed.

At this point the final phase-shift occurs with which the Ice Age begins.

Once the Primer Fields have vanished we are only half-way through Phase 1 of the climate collapse

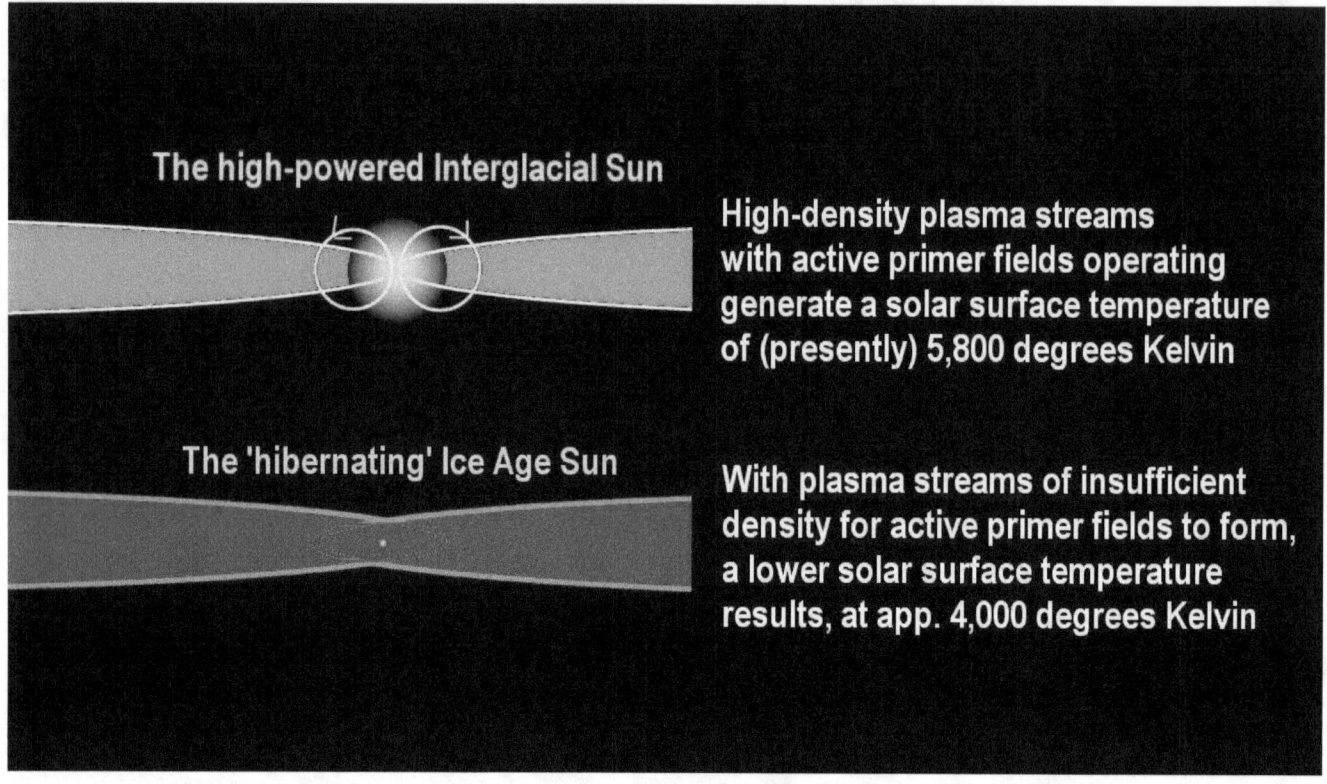

Once the Primer Fields have vanished, concentrated plasma is no longer focused onto the Sun. The interstellar plasma stream then simply flows by the Sun, loosely focused, with a default-type of density that keeps the Sun operating in a low-power mode with a potential surface temperate in the 4,000-degrees range, instead of the current 5,800 degrees Kelvin.

This means, in practical terms, these huge events are already happening now, with the world getting ever-colder and drier, year after year, while we are only half-way through Phase 1 of the climate collapse, with a lot more yet to come.

Ever-larger temperature extremes as the greenhouse is diminishing

The moderating greenhouse effect of the atmosphere narrows the cosmic temperature extremes to a nicely liveable climate.

Greenhouse effect produced by water vapor in the atmosphere

without the greenhouse effect of the Earth's atmosphere:
night temperature -170 decrees C
day temperature +117 degrees

Earth's greenhouse effect is diminished by cosmic-ray increase

cloud nucleation reduces water vapor: deeper droughts and lesser greenhouse

other greenhouse contributions
CO2 greenhouse contribution

cosmic-rays increase cloud nucleation

We also see ever-larger temperature extremes happening as the atmosphere's greenhouse effect is diminishing by increasing cloud nucleation that takes evermore water vapour, which is the main greenhouse gas, out of the atmosphere.

The entire climate-change process by solar cosmic-ray flux

The entire climate-change process, as I had noted earlier, is driven by increasing solar cosmic-ray flux that enhances cloud nucleation. Increased cloudiness makes the Earth colder, because the white tops of clouds reflect a portion of the incoming solar radiation back into space.

Cosmic-ray flux is continuously increasing

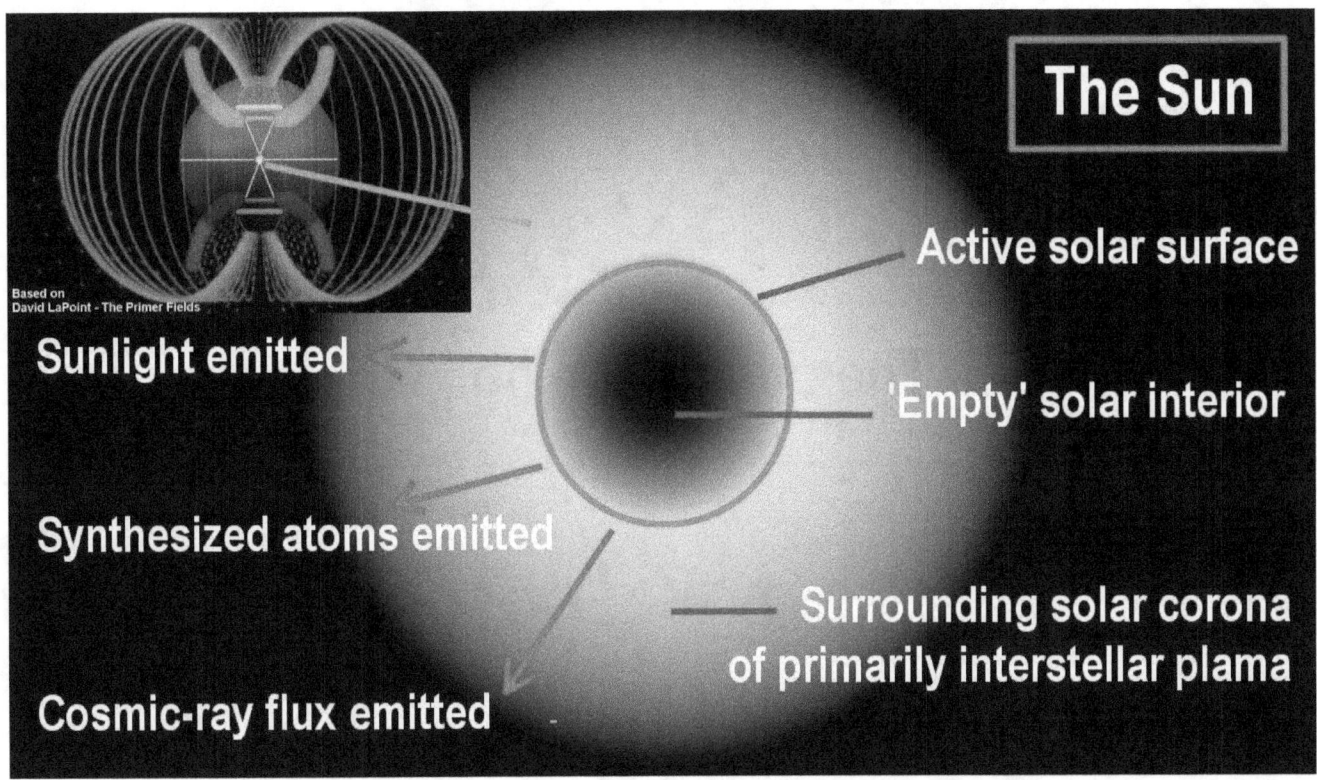

Cosmic-ray flux is continuously increasing all the way through the boundary time-zone, because solar activity is diminishing. A weaker Sun has a weaker plasma corona surrounding it, where a portion of the solar cosmic-ray flux is trapped. When the corona diminishes, cosmic-ray flux is increasing.

The critical timing

The critical timing

The critical timing

Throughout Phase I in the boundary zone

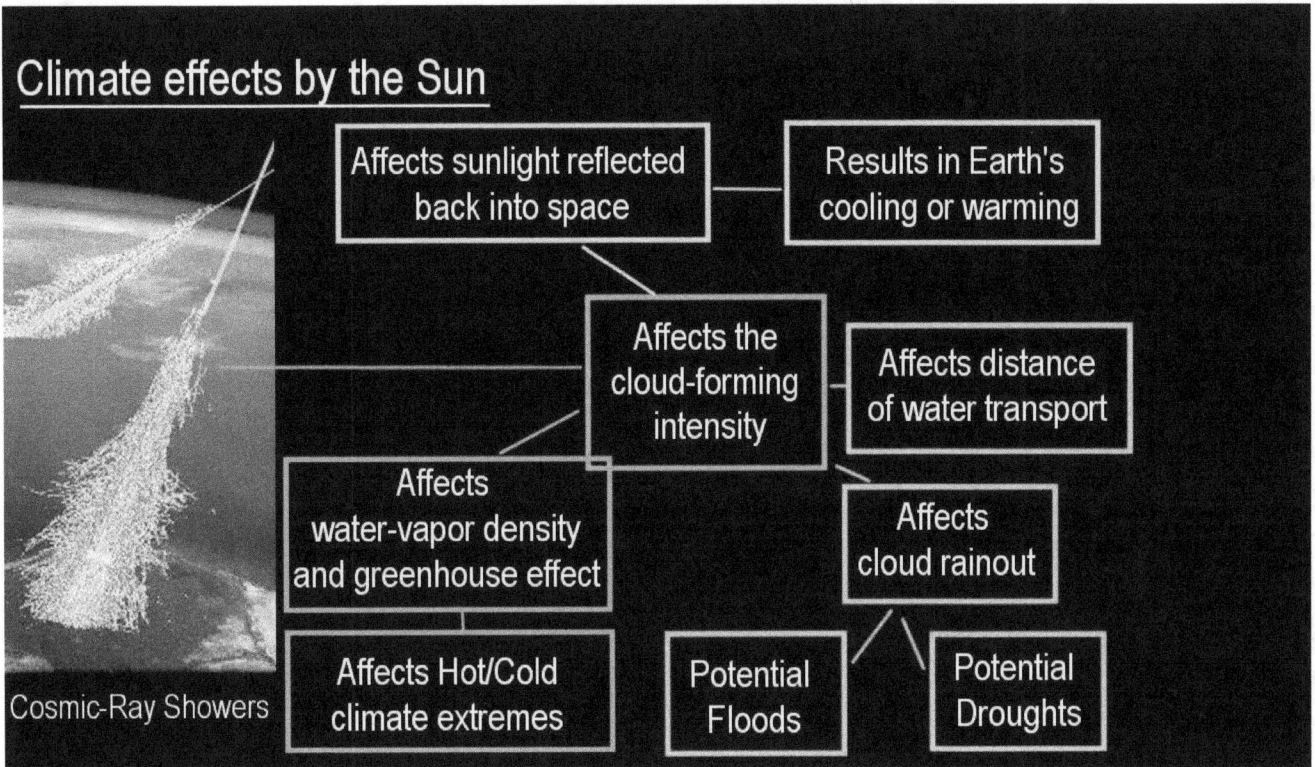

Throughout Phase I in the boundary zone towards the next Ice Age, all the big climate-change effects on Earth, are, and will be, caused by changing solar cosmic-ray flux that is affecting our atmosphere. That's the current link between the Sun and our climate.

Solar light and heat radiation nearly constant

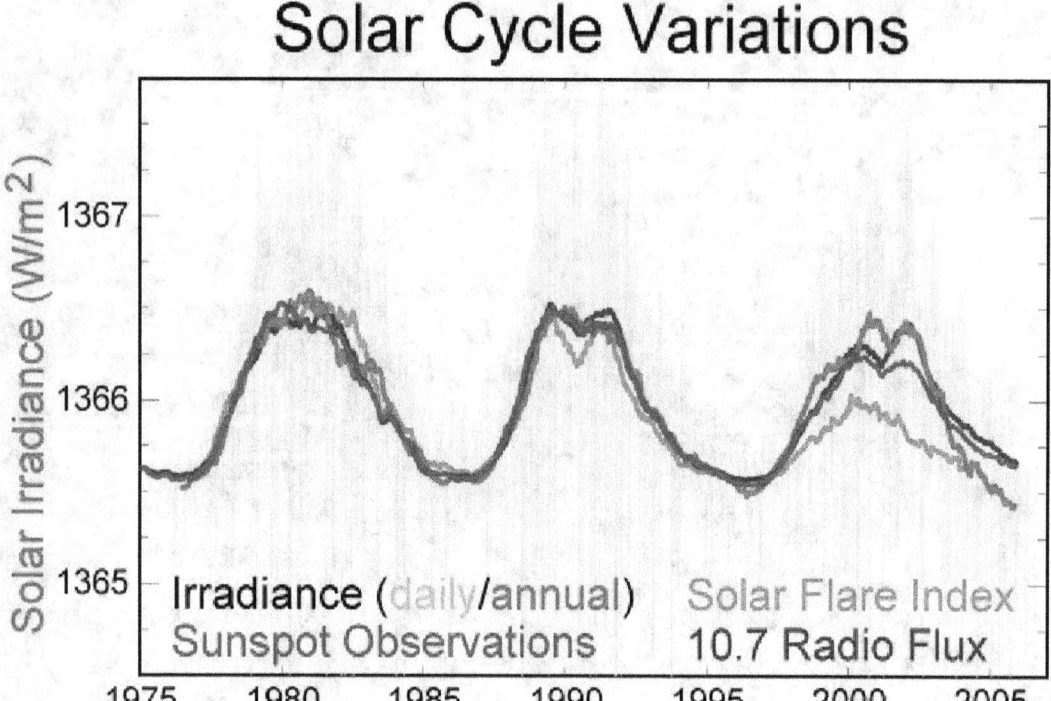

The solar light and heat radiation remains nearly constant, as I said before, within a fraction of percent. This continues all the way through Phase 1 of the diminishing solar dynamics, while the Earth is getting colder and colder by cosmic-ray effects.

Only when we get into Phase 2

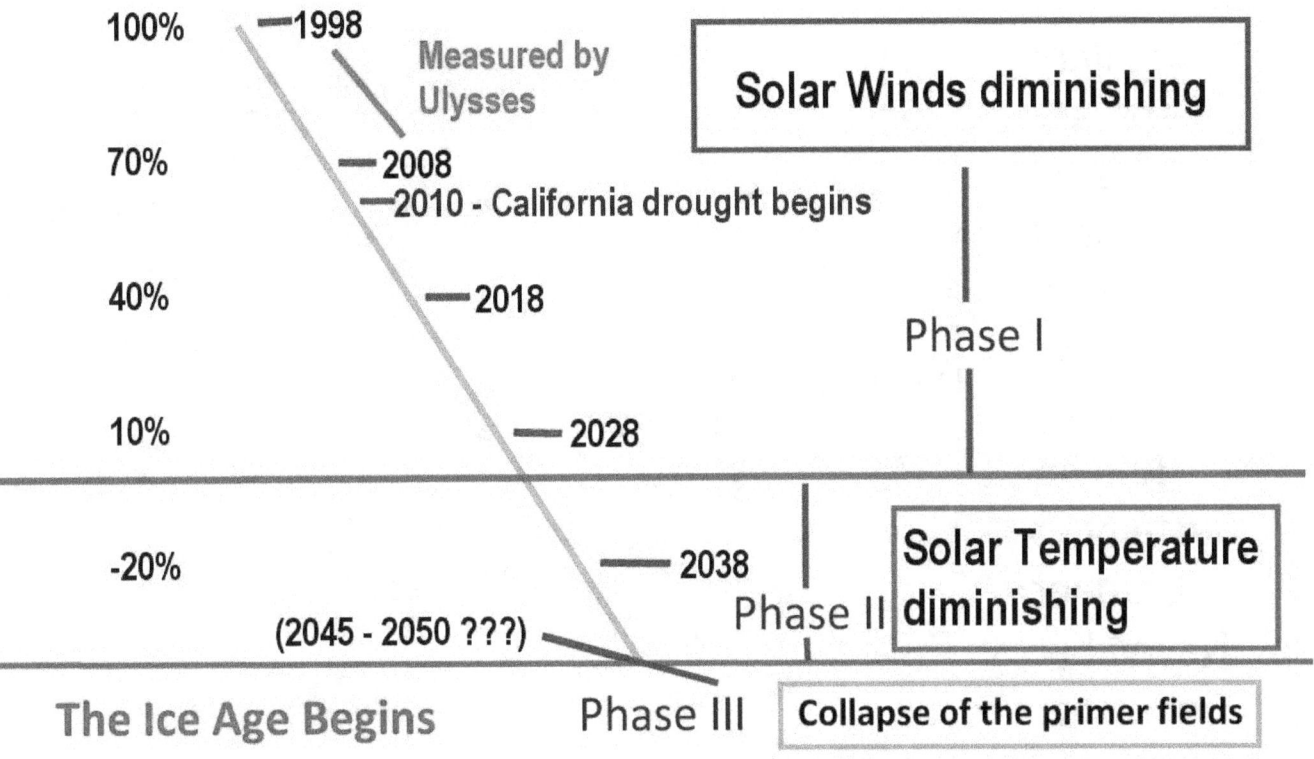

Only when we get into Phase 2, will the Sun's radiation begin to diminish. It will diminish in addition to the continuously increasing cosmic-ray flux.

It is highly likely that agriculture will collapse before we get to this point, while we are still in Phase 1, or it will happen early in Phase 2.

The available growing season is barely sufficient

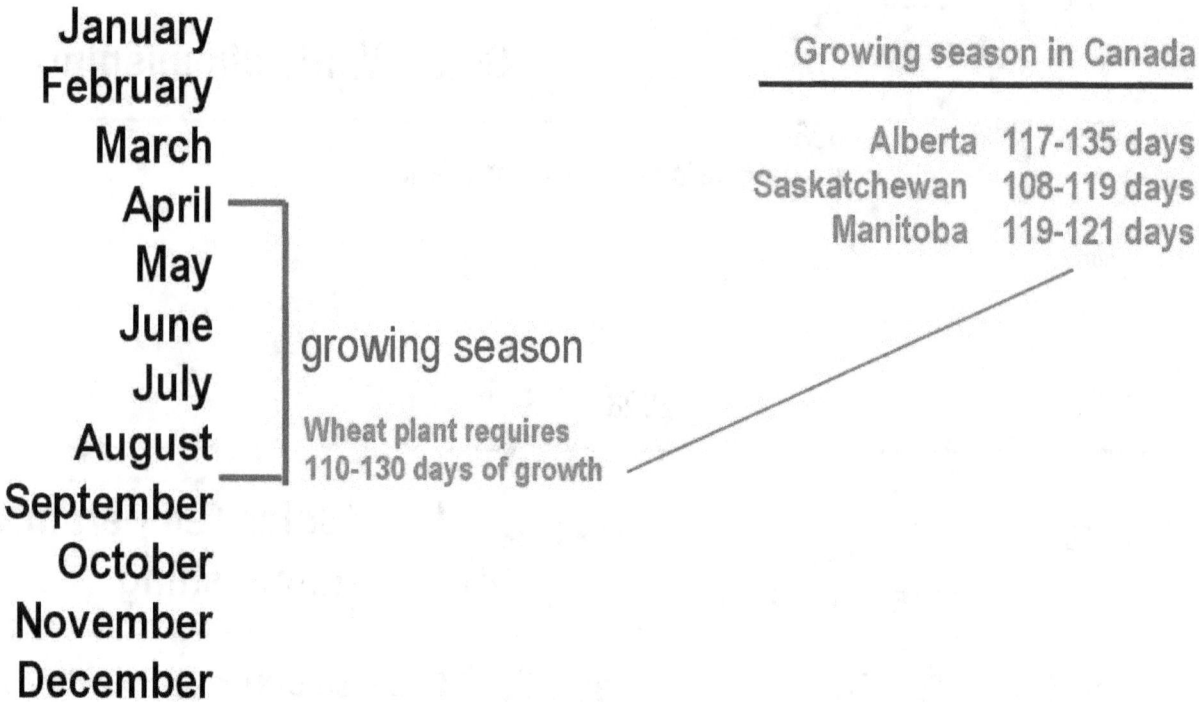

In most regions, like in Canada, the available growing season, for food production, is barely sufficient for crops to be planted, to grow and mature, and to be harvested.

Large areas will become unsuitable for food production

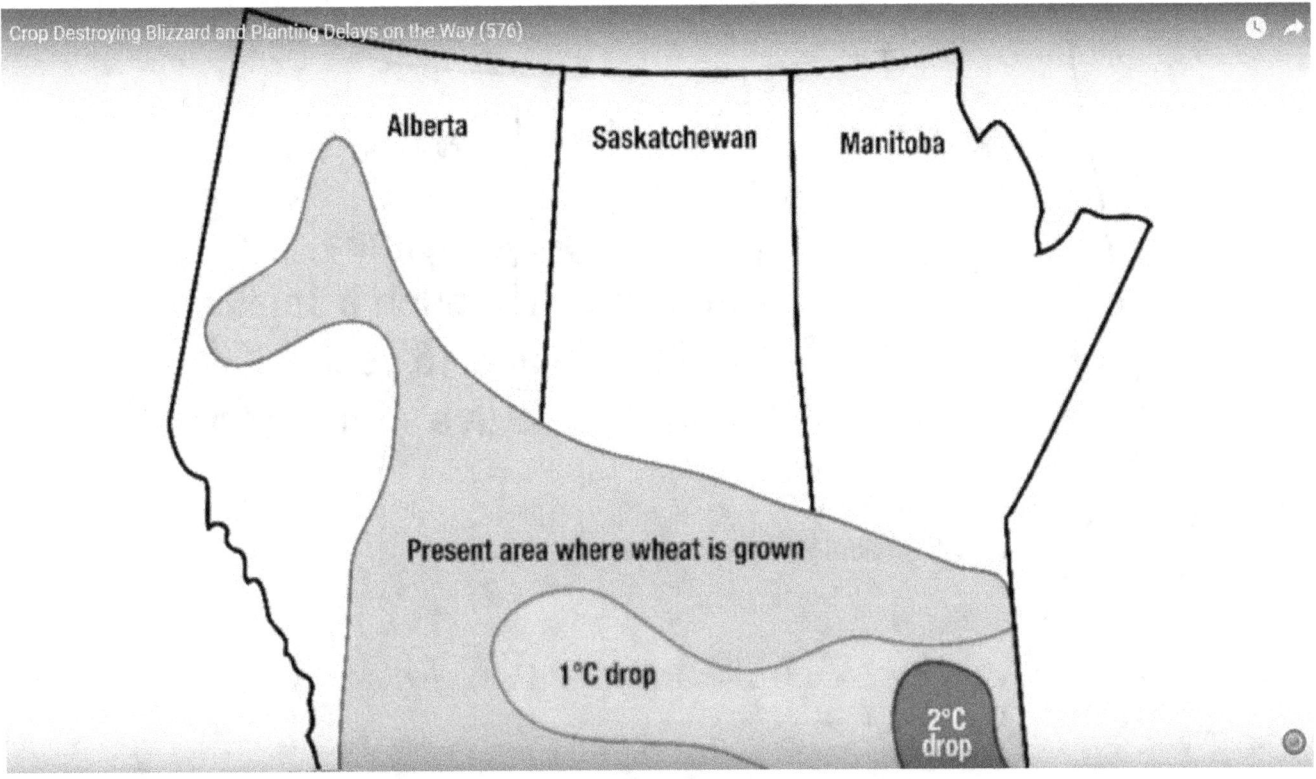

When the increasingly colder climate that is now developing, shrinks the growing window in the spring and in the fall, large areas will become unsuitable for food production.

Could shrink the food growing area in Canada's grain belt

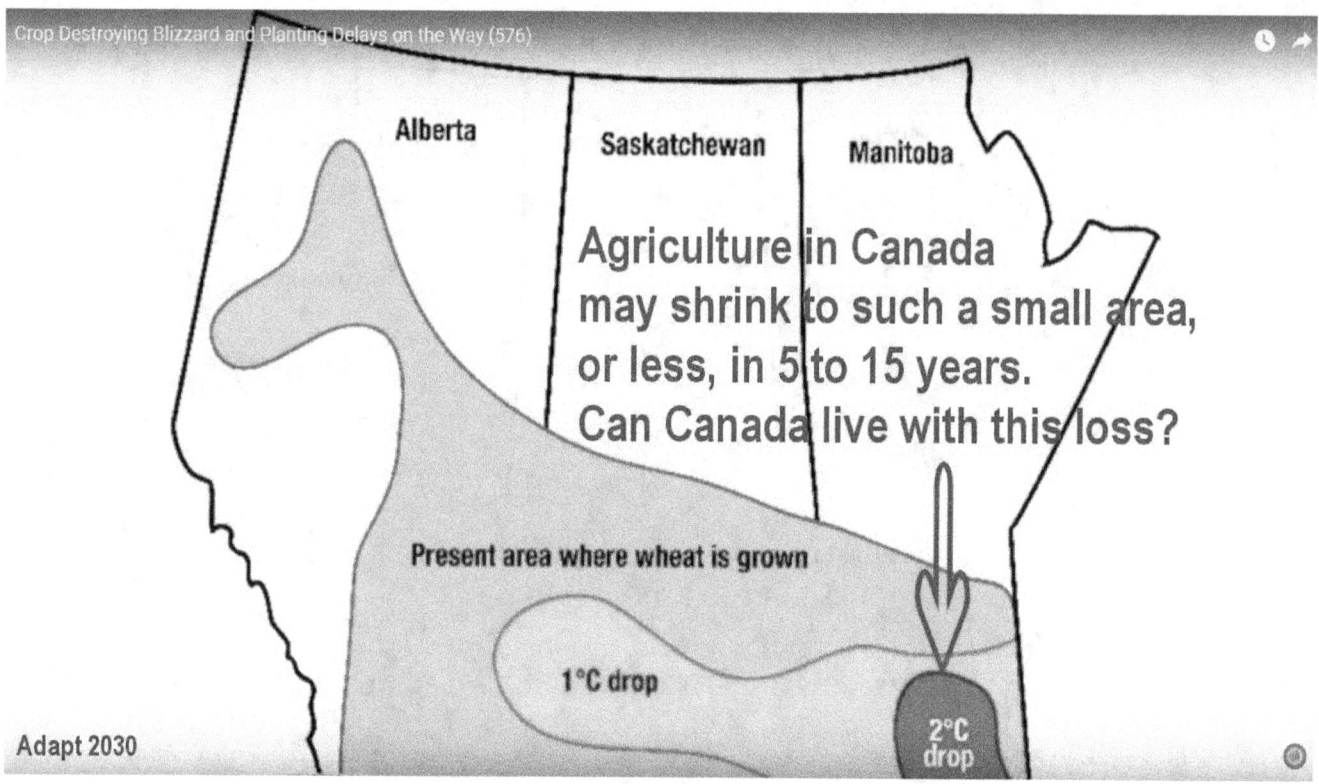

According to a study, a hypothetical two-degrees drop in average temperature, could shrink the food growing area in Canada's grain belt to just a few percent of its present size. Nor does this type of study apply only to Canada. Most food growing regions appear to be similarly vulnerable to the now increasingly colder climates.

The Ice Age Challenge

is not of a type that can be responded to reactively,

like Nuclear War.

If we delay our response till it happens, it will be too late.

The Ice Age Challenge

is not of a type that can be responded to reactively,

like Nuclear War.

If we delay our response till it happens, it will be too late.

What happens after Phase 1 is of no great significance

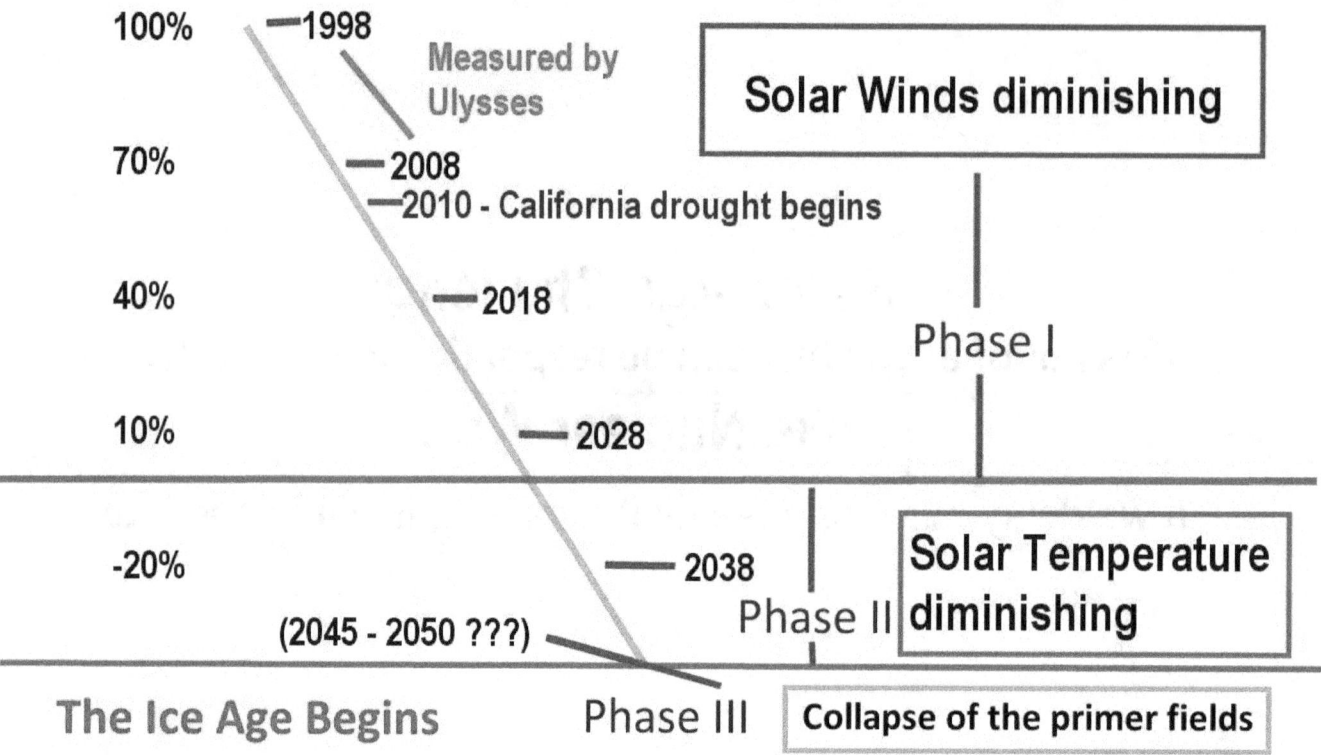

Ironically, what happens after the end of Phase 1 is actually of no great significance to humanity. That's almost a paradox.

What happens during Phase 2, which leads towards the Primer Fields collapsing, and to the start of Phase 3 that becomes the full Ice Age, should logically be main concern. Ironically it isn't, because it won't affect humanity in any big way.

This is so, because if humanity fails itself in the present, and this includes all of us, by us not building for us the critical New World with indoors agriculture as a fall-back cushion, then there won't be many people remaining alive when Phase 3 starts. In this case it won't matter, whether the phase-shift to the next Ice Age happens in the 2040s, or the 2050s, or in the 2060s, because almost nobody would be around then, to witness the event.

Inversely, if we, all of us working together, were to succeed in building the New World infrastructures that make our food production climate independent, then it wouldn't matter either, whether Phase 3 begins in the 2040s, the 2050s, or the 2060s, because then, by us living securely in our protected New World, the start-up of the next Ice Age wouldn't affect us. It would then be just a curiosity.

The Universe won't let us wait

> **Fortunately the Universe won't let us wait till it is too late.
> The fringe effects of the weakening Sun-system impel us already
> into types of actions with which the Ice Age Challenge can be met.**

The Universe won't let us wait

fringe effects impel us already into actions that can meet the Ice Age Challenge

The Ice Age start-up will happen

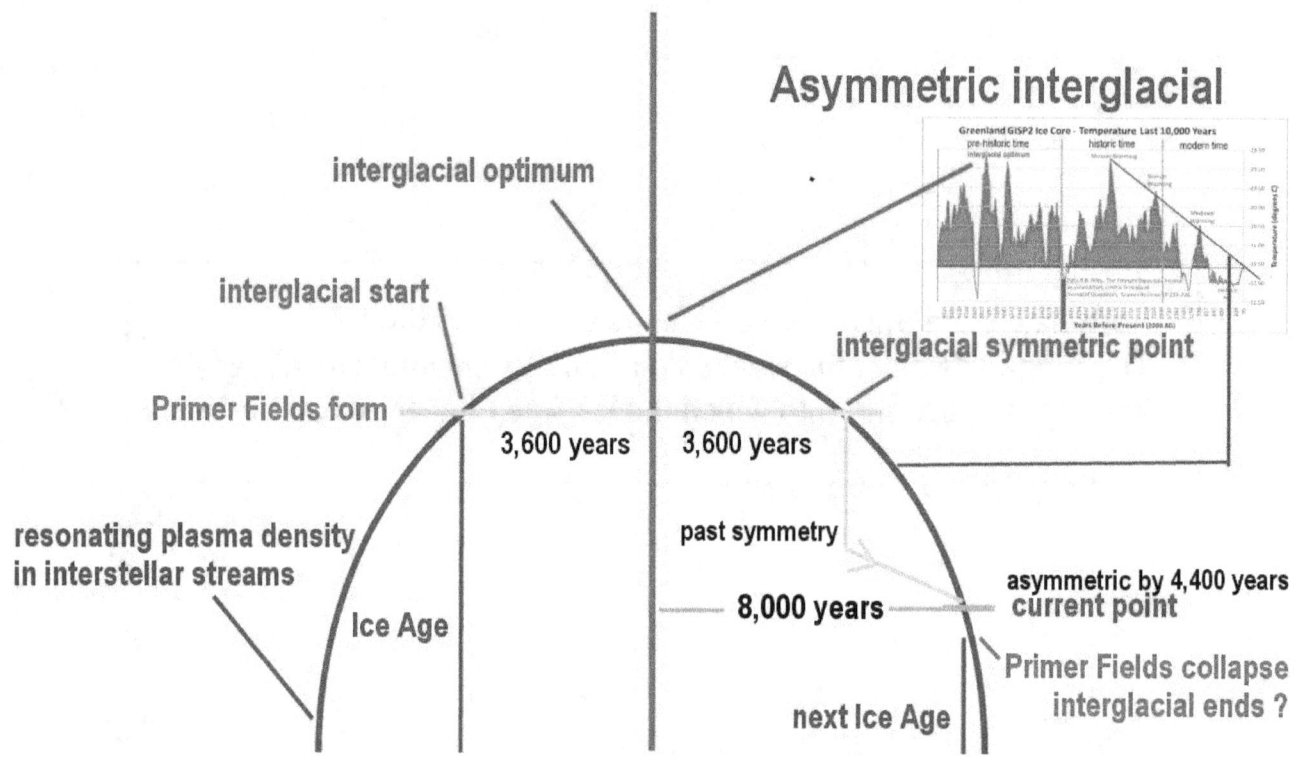

One way or another the Ice Age start-up will happen. Only we will be affected by our action or inaction. The Primer Fields will collapse. Our precious interglacial warm holiday from the Ice Age landscape, will end. That's inevitable. The end of the interglacial warm period is the result of a long-term 120,000-years electric resonance in the interstellar plasma stream that our solar system is a node point of.

As it is, the interglacial warm period only happens near the very peak of this resonance pulse. That's when the plasma density in the interstellar stream exceeds the start-up threshold for the Primer Fields to form. In the present case, the start-up occurred only 3,600 years before the maximum of the resonance pulse.

We are presently 8,000 years past the maximum level of the interstellar resonance pulse. This places us on a more-steeply diminishing slope.

We are on this steeper slope, because the Primer fields are highly resilient. This resilience has puts us 4,400 years past the density level at which the Primer Fields had been established. The reason for this delay is, that it takes a far greater start-up density for the Primer Fields to form, than its required to maintain them. They hang on longer, because of a built-in positive feed-back in their operation. For how much longer, however, they will remain in operation cannot be forecast.

The global warming pulses have been getting smaller

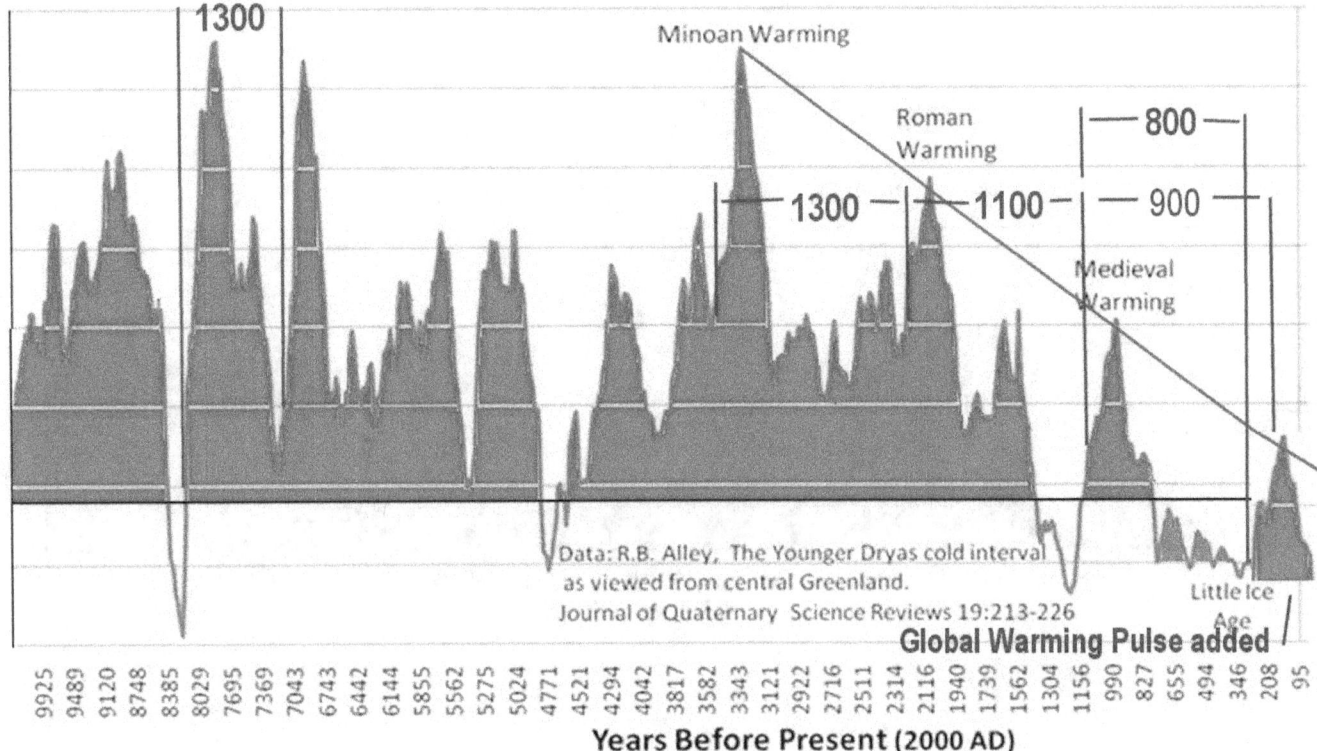

The evidence that we are way past the symmetry point, is found in ice core samples. Over the last 3,500 years the global warming pulses have been getting smaller and smaller.

The global warming pulses are the result of resonance effects within the structure of the solar system itself. Thus, with the interstellar plasma stream now diminishing evermore rapidly, the resonance effects diminished likewise, and occur with ever-shorter intervals between them.

The short-term oscillations too, have diminished

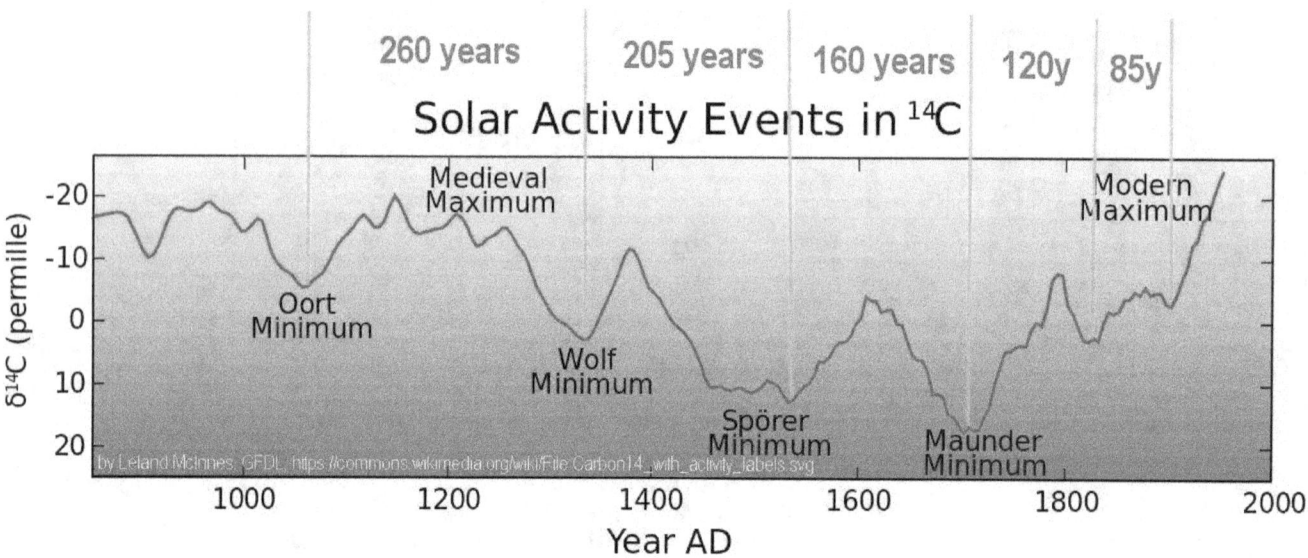

The short-term oscillations too, of the last 1,000 years, have diminished in the same manner. These are the short term oscillations that gave us the grand solar minimum periods, from the Oort Mini the end of Phase 1 is actually of no great significance to humanity. That's almost a paradox.

What happens during Phase 2, which leads towards the Primer Fields collapsing, and to the start of Phase 3 that becomes the full Ice Age, should logically be main concern. Ironically it isn't, because it won't affect humanity in any big way.

This is so, because if humanity fails itself in the present, and this includes all of us, by us not building for us the critical New World with indoors agriculture as a fall-back cushion, then there won't be many people remaining alive when Phase 3 starts. In this case it won't matter, whether the phase-shift to the next Ice Age happens in the 2040s, or the 2050s, or in the 2060s, because almost nobody would be around then, to witness the event.

Inversely, if we, all of us working together, were to succeed in building the New World infrastructures that make our food production climate independent, then it wouldn't matter either, whether Phase 3 begins in the 2040s, the 2050s, or the 2060s, because then, by us living securely in our protected New World, the start-up of the next Ice Age wouldn't affect us. It would then be just a curiosity.

The Universe won't let us wait

> Fortunately the Universe won't let us wait till it is too late.
> The fringe effects of the weakening Sun-system impel us already into types of actions with which the Ice Age Challenge can be met.

The Universe won't let us wait

fringe effects impel us already into actions that can meet the Ice Age Challenge

The Ice Age start-up will happen

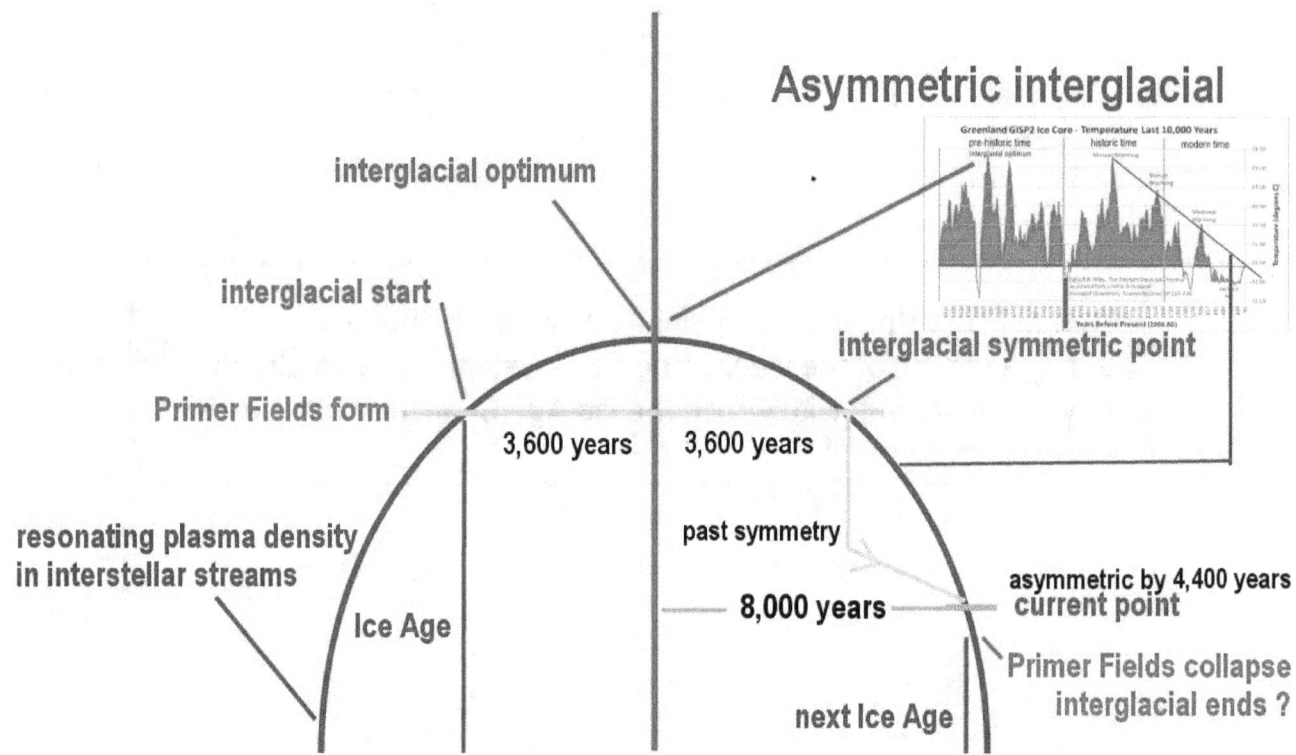

One way or another the Ice Age start-up will happen. Only we will be affected by our action or inaction. The Primer Fields will collapse. Our precious interglacial warm holiday from the Ice Age landscape, will end. That's inevitable. The end of the interglacial warm period is the result of a long-term 120,000-years electric resonance in the interstellar plasma stream that our solar system is a node point of.

As it is, the interglacial warm period only happens near the very peak of this resonance pulse. That's when the plasma density in the interstellar stream exceeds the start-up threshold for the Primer Fields to form. In the present case, the start-up occurred only 3,600 years before the maximum of the resonance pulse.

We are presently 8,000 years past the maximum level of the interstellar resonance pulse. This places us on a more-steeply diminishing slope.

We are on this steeper slope, because the Primer fields are highly resilient. This resilience has puts us 4,400 years past the density level at which the Primer Fields had been established. The reason for this delay is, that it takes a far greater start-up density for the Primer Fields to form, than its required to maintain them. They hang on longer, because of a built-in positive feed-back in their operation. For how much longer, however, they will remain in operation cannot be forecast.

The global warming pulses have been getting smaller

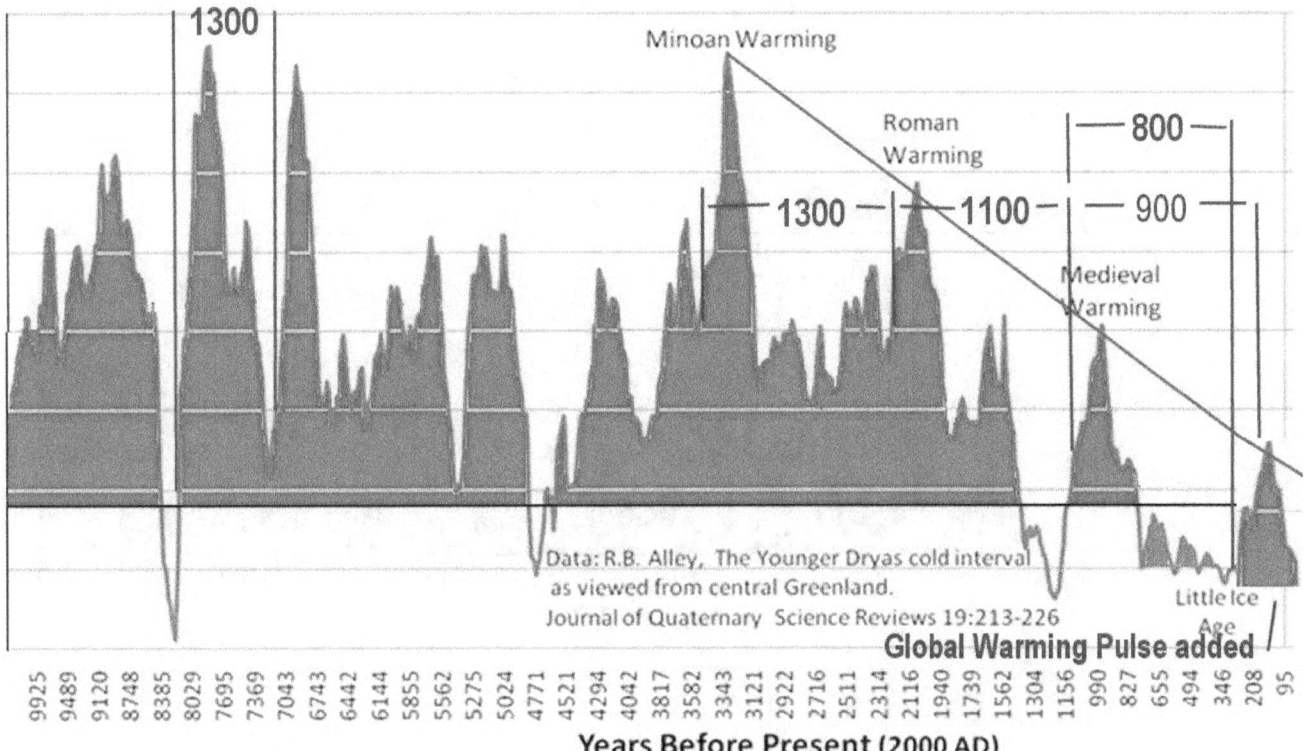

The evidence that we are way past the symmetry point, is found in ice core samples. Over the last 3,500 years the global warming pulses have been getting smaller and smaller.

The global warming pulses are the result of resonance effects within the structure of the solar system itself. Thus, with the interstellar plasma stream now diminishing evermore rapidly, the resonance effects diminished likewise, and occur with ever-shorter intervals between them.

The short-term oscillations too, have diminished

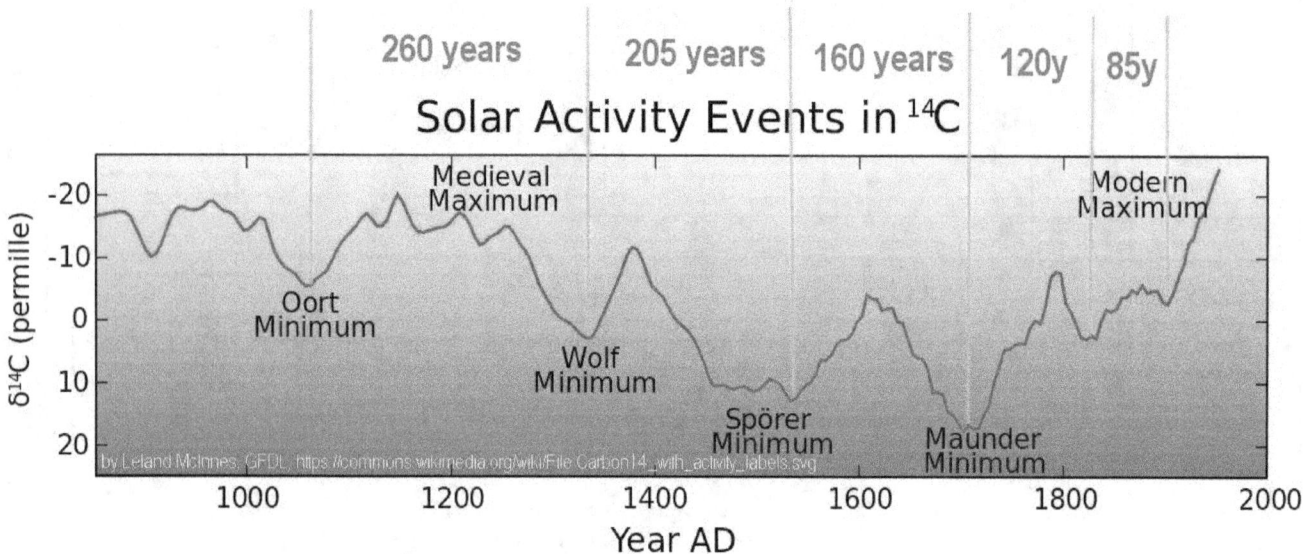

The short-term oscillations too, of the last 1,000 years, have diminished in the same manner. These are the short term oscillations that gave us the grand solar minimum periods, from the Oort Minimum to the Maunder Minimum and beyond. They too, have diminished dramatically, both in amplitude and in cycle times.

Diminished in a near-geometric progression

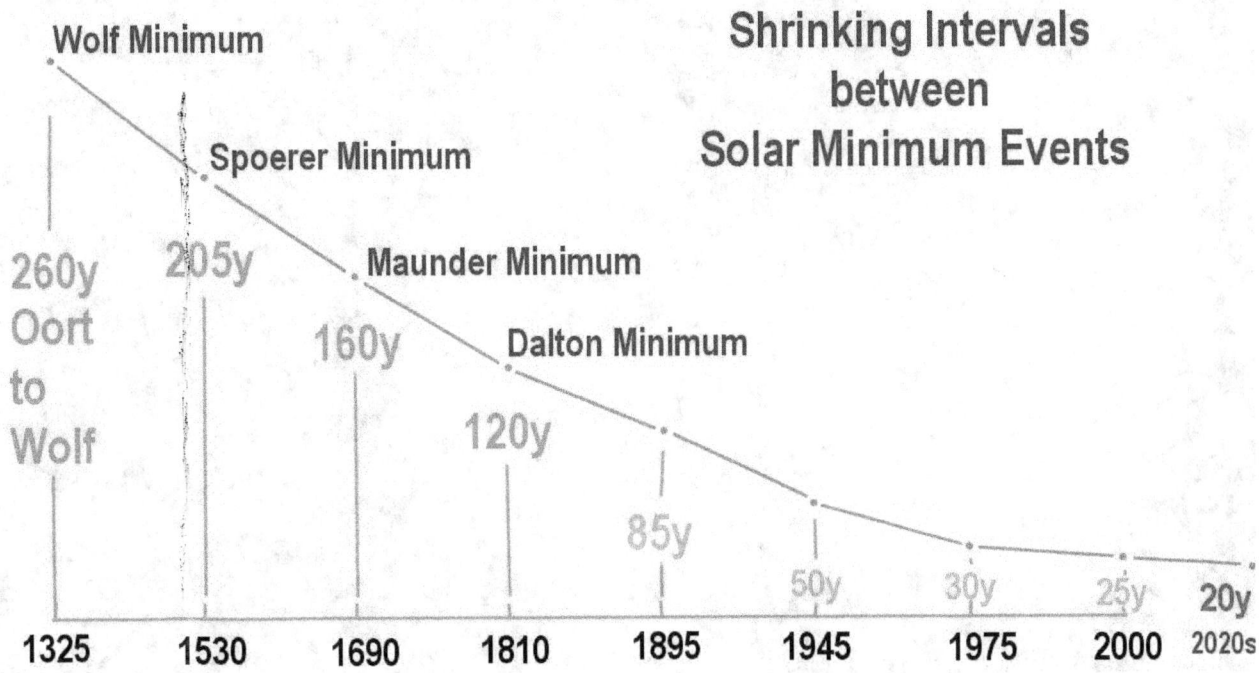

These short-term cycles have diminished in a near-geometric progression, to the point of insignificance. They are so minuscule now that they won't affect anything anymore.

This is not something that may happen in the future. It is happening now.

The solar heartbeat is slowing

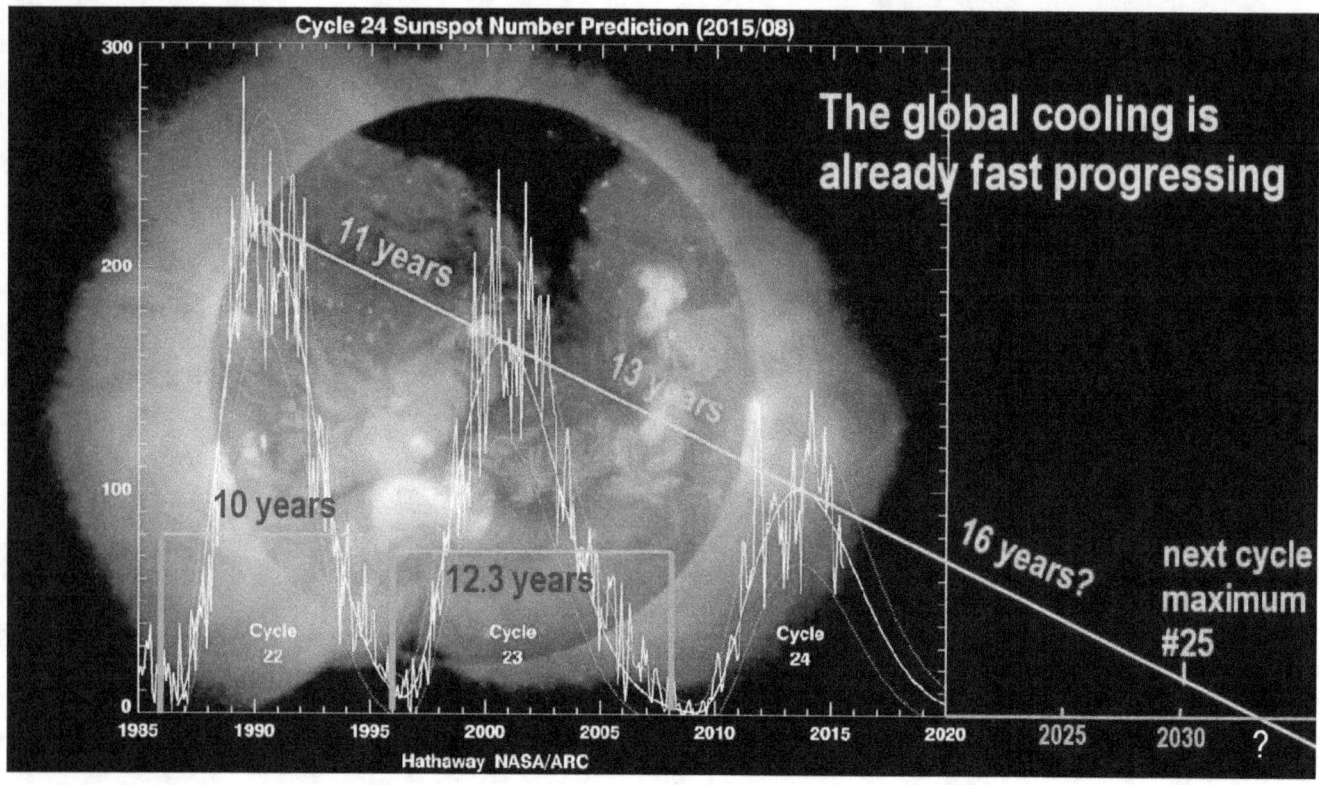

Even the solar cycles are diminishing, and their heartbeat is slowing, while at the same time the cosmic-ray flux is increasing.

Every vital measurement seems to warn us that we are getting close to the end point at which the Primer Fields break down and the final phase shift to glaciation happens. This means reverting back to what is termed the Ice Age, which had been the climate norm for the Earth for the last half a million years.

The electric resonance gives the Earth its brief interglacial holidays

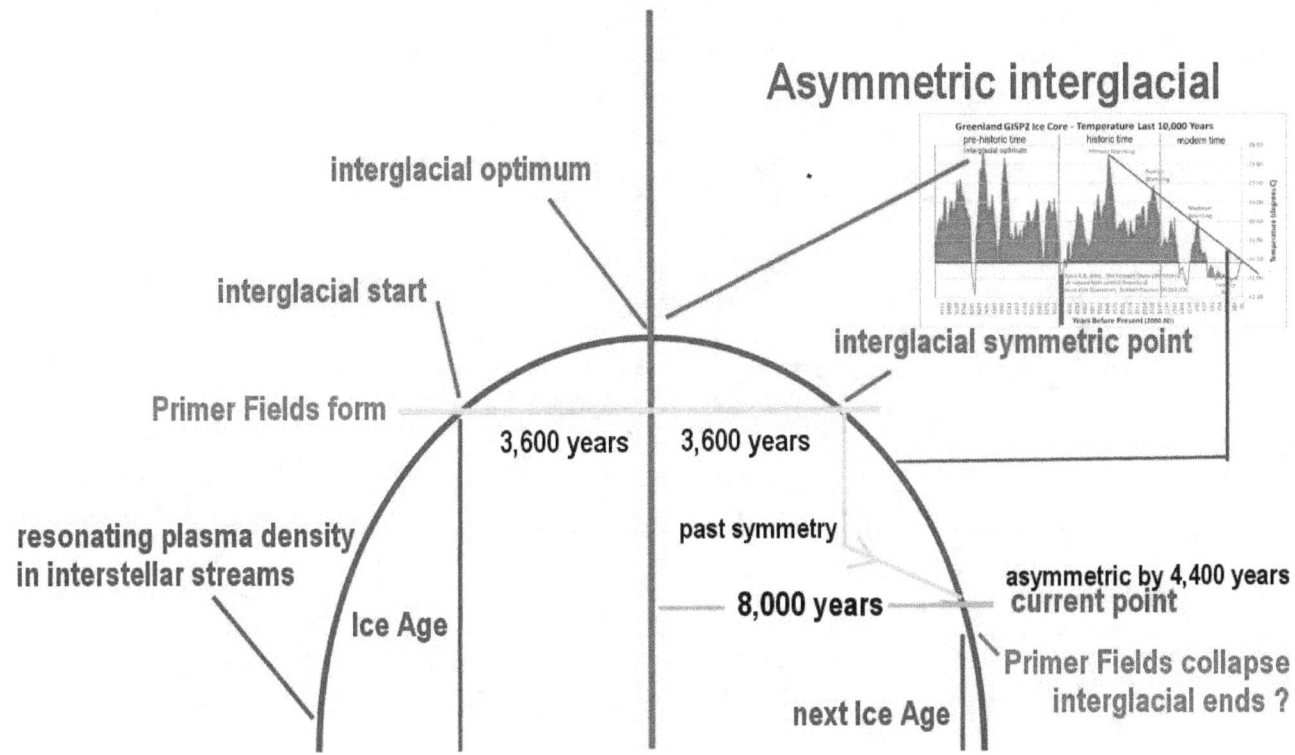

Actually, it is only because of the natural electric resonance in the in the interstellar plasma stream, that conditions occur every 120,000 years, which bring the plasma density up to the start-up level for the primer fields. So it is the electric resonance effect that gives the Earth its brief interglacial holidays from the cold.

If the resonance effects didn't happen none of us would exist

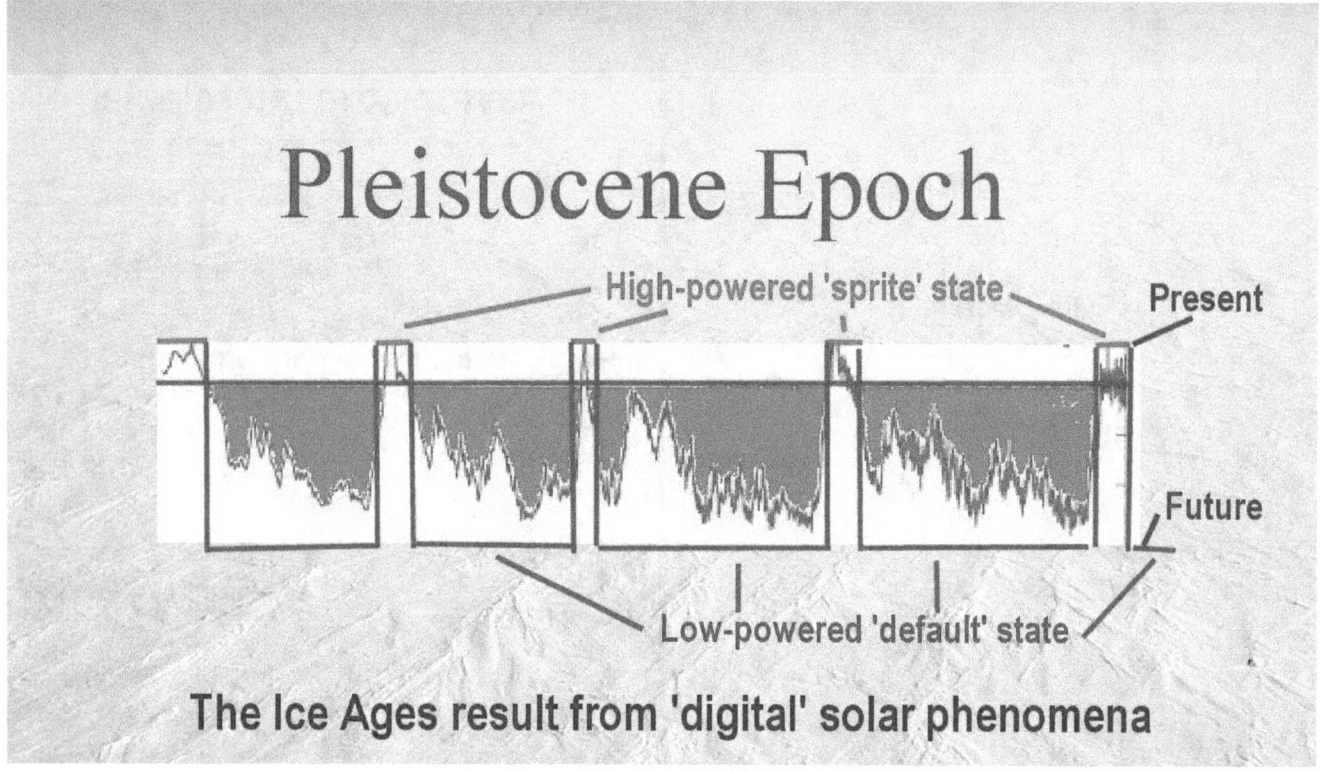

If the resonance effects didn't happen, the interglacial holydays wouldn't happen either, and none of us would exist.

Fortunately, the interglacial holidays do happen. At the current stage, they have enabled us to develop ourselves technologically to the point that for the first time ever, in all of human history, we have the power developed to build us a type of New World that enables us to ride into the next Ice Age, undiminished, with a 7-billion world population, and to flourish in the harsh glaciation climate, which, as I said, is the normal climate in the Pleistocene Ice Age epoch.

If we fail to live up to what we have become capable of as human beings, we will drop off the interglacial cliff and fall into oblivion.

However, should we succeed to live as human beings, we will still fall off the interglacial cliff, but we will have a soft landing prepared for us, on a pillow built with the resources that we have within us as human beings.

To get there, we need to step away from the perversion of science that H. G. Wells had inspired with his novel the Time Machine.

Whether we will lift this science-chokehold off us, and begin live again and have a bright future, will be answered by us all in the near term. If we fail, we fail collectively. If we succeed we succeed collectively. Individual exceptions are not possible in principle. For this we need to ask ourselves individually, am I doing enough to assure that we will succeed collectively and live? That's the most basic question that should be asked. And bigger questions than this, need also to be asked.

Humanity doesn't exist to merely live

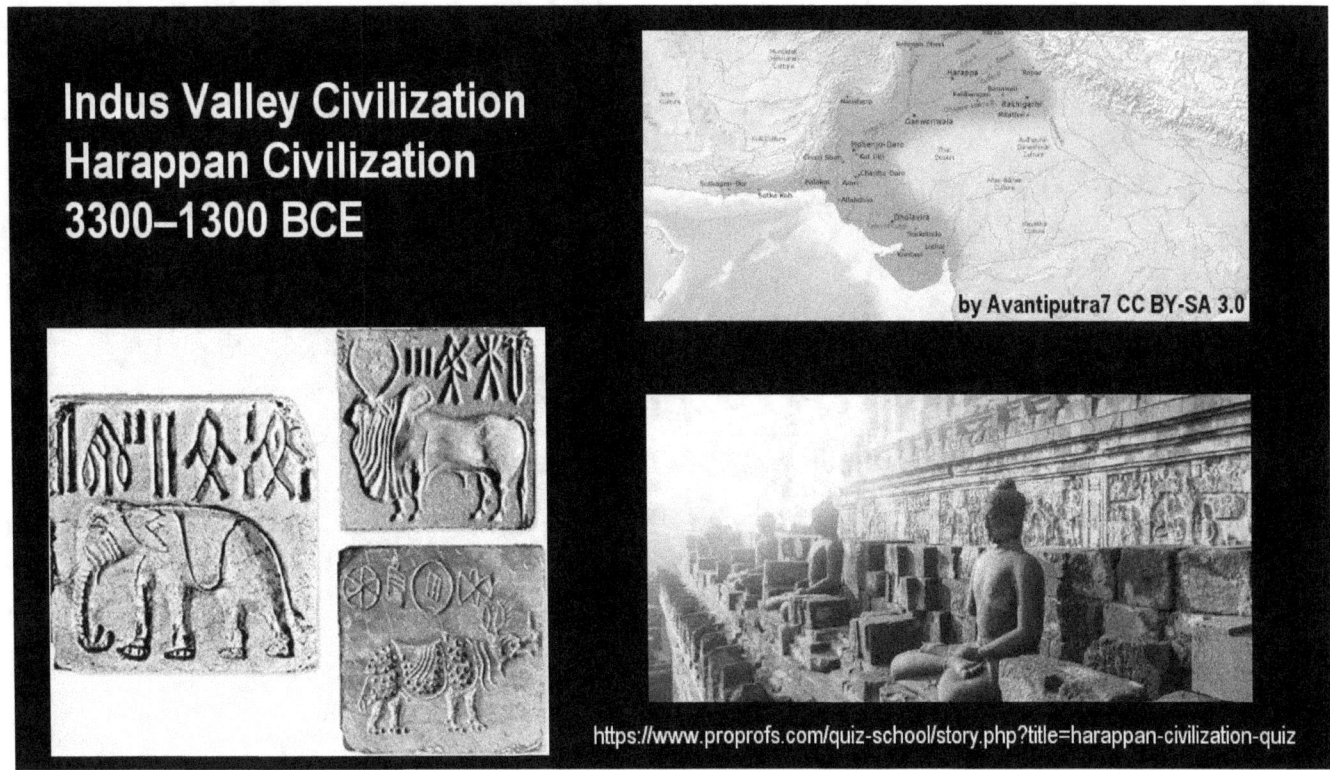

Humanity doesn't exist to merely live.

To be satisfied with merely living is like keeping a car parked in the garage as the optimum goal. A car is designed for diving, for freedom in transportation, not to be in a garage. Humanity too, is designed not just to exist, but to unfold with ever-greater scientific, productive, cultural, and creative expressions that unfold the human potential like a rose is unfolding its beauty in the sunshine. The potential in human living is to create an evermore powerful, richer, and more-beautiful world in which the name, Humanity, is defined, and is defined as the brightest gem of the Earth.

The New World infrastructures cannot be build on a platform of failures

The Christian Martyrs' Last Prayer, by Jean-Léon Gérôme (1883). Roman Colosseum.

The imperial paradigms of ego, greed, private monetarism, manipulation, geopolitics, financial looting, economic colonialism, warfare old and new, including the subjugation of nations to poverty and slavery, are not hallmarks of human civilization. They are examples of failures, on the path of growing up. The New World infrastructures cannot be build on a platform of failures. The modern necromancy of money, and the isolation it creates in society, including inhumanity and poverty, are unfit foundations for building anything on, much less a new world.

The only foundation on which we can succeed, is the power of our humanity

The only foundation on which we can succeed, is the power of our humanity, our love and care for one another, and our ability to supersede all limits, with the joy to see them disappear.

Our historic achievements should be seen as a foundation to build on

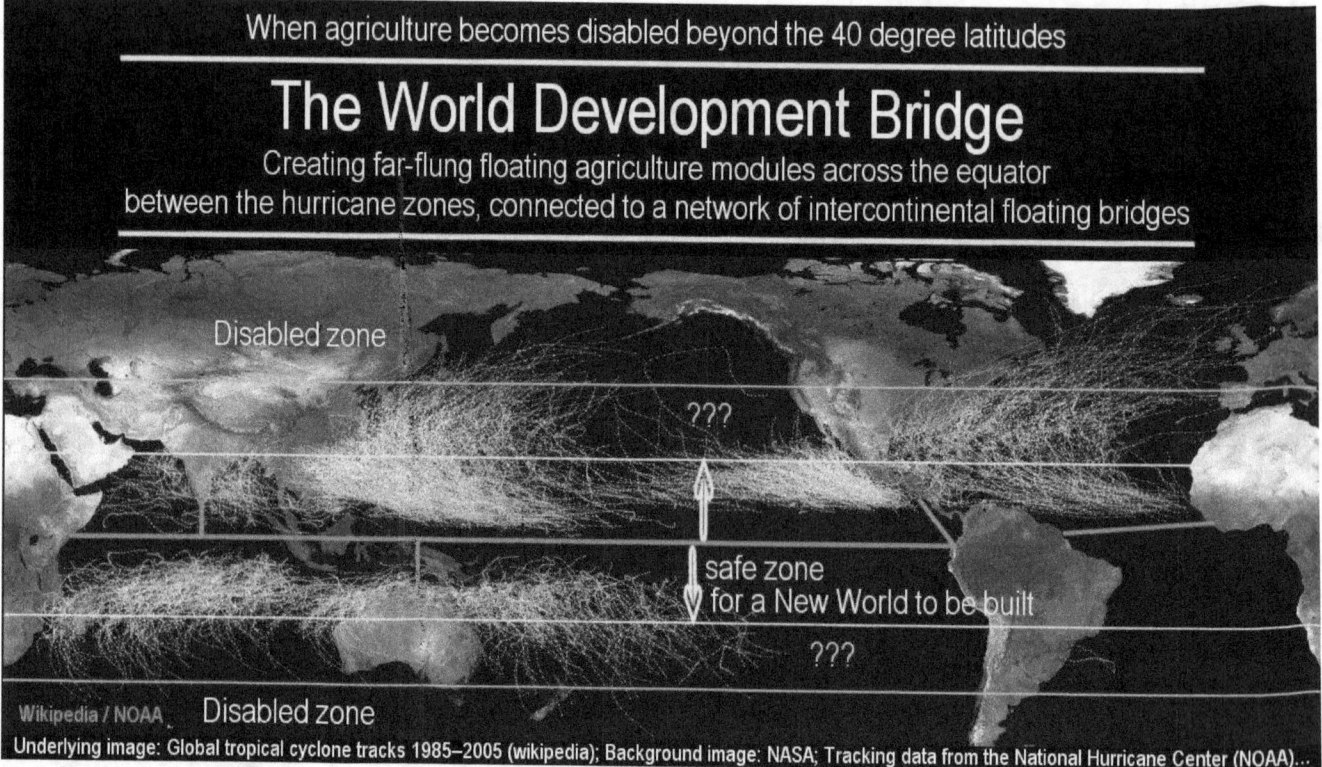

Our building us a new world, afloat across the equatorial sea, should not be seen as merely the most efficient way to meet an urgent need, but should be seen a 'canvass' on which the superlatives of our humanity are becoming ever-further revealed in expression, drawing the cultures of the world together into a richly varied human tapestry.

The goal should be, to not merely preserve the cultures of civilizations that have developed over millennia, each by their own human resources, - such as in the Tigris and the Euphrates regions of the old world; and in Egypt along the Nile; and in India along the Indus and the Ganges; and in China along the Yellow River and the Yangtze, and so forth. These historic cultures have coloured the human landscape with traditions and achievements with which humanity has enriched each other, out of which the present humanity has emerged. But this emergence from the past should not lead to a dead-end in the future with most of humanity being poised for oblivion, as the ice age promises. Our historic achievements should be seen as a foundation to build on, to build a New Word with evermore advanced economics, truthfulness in science, and technologies that have produced evermore useful wonders for ever-greater freedoms from limitations of all sorts.

If this is the direction in which we seek our future

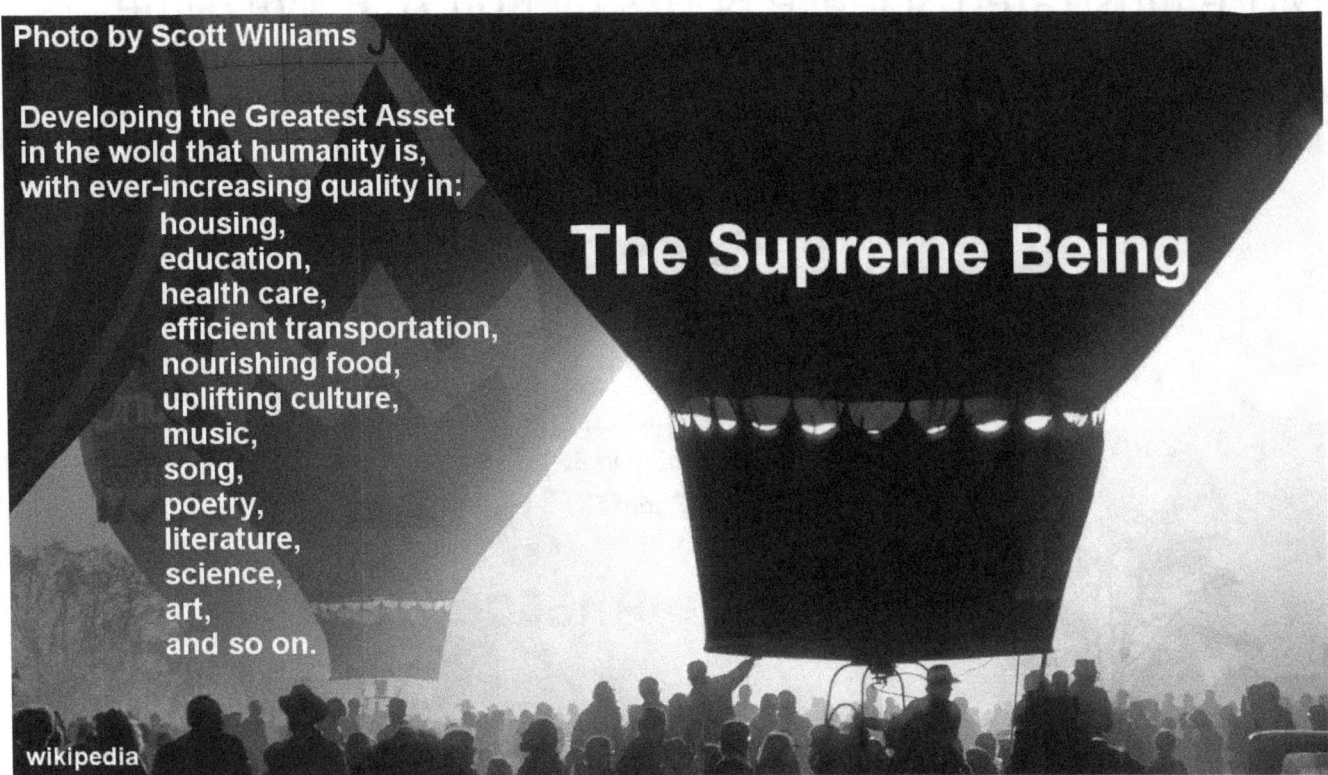

If this is the direction in which we seek our future, the Ice Age Challenge that is now on the table as a huge challenge, will drift into the background to be met in stride on the road to ever greater achievements, of which it may be said that we haven't seen anything yet.

More Illustrated Science Books by Rolf A. F. Witzsche